P9-ARU-826

Experimental Sound & Radio

TDR Books

Richard Schechner, series editor

Puppets, Masks, and Performing Objects, edited by John Bell
Experimental Sound & Radio, edited by Allen S. Weiss

Experimental Sound & Radio

Edited by Allen S. Weiss

A TDR Book

The MIT Press
Cambridge, Massachusetts
London, England

The Drama Review editorial staff for the "Experimental Sound & Radio" issue:
Associate Editor: Mariellen R. Sanford
Managing Editor: Marta Ulvaeus
Assistant Editor: Julia Whitworth
Editorial Assistant: Christine Dotterweich

Previously published as a special issue of *The Drama Review* (Vol. 40, no. 3, Fall 1996).

Library of Congress Cataloging-in-Publication Data

Experimental sound & radio / edited by Allen S. Weiss.
p. cm.—(TDR Books)
"Originally published as a special issue of The drama review, vol. 40, no. 3, fall 1996"—P.
"A TDR book."
Includes bibliographical references and index.
ISBN 0-262-73130-4 (pbk.:alk. paper)
1. Experimental radio programs—History and criticism. I. Title: Experimental sound and radio. II. Weiss, Allen S., 1953– III. Drama review. IV. Series

PN1991.8.E94 E97 2001
791.44'6—dc21 00-068700

Contents

Radio Icons, Short Circuits, Deep Schisms

Allen S. Weiss

Multiple and contradictory histories of radiophony could be constituted, depending upon both the historical paradigms chosen to guide the research, and the theoretical phantasms behind the investigation. Its prehistory is vast; one key moment may be cited. Rabelais, in the fourth book of *Pantagruel*, describes a seafaring voyage during which the crew hears voices that seem to come from thin air, an effect causing great fear. Pantagruel explains that these sounds consist of words that were frozen in the winter air, and which begin to thaw out upon being touched, thus becoming audible.[1]

> And we could see sharp words, bloody words (which, according to the pilot, sometimes went back to the place where they'd been spoken, only to find the throat that uttered them had been slit open), horrible words, and many others equally unpleasant to see. And when they'd melted, we heard: *hin, hin, hin, hin, hiss, tick, tock, whizz, gibber, jabber, frr, frrr, frrr, boo, boo, boo, boo, boo, boo, boo, boo, crack, track, trr, trr, trr, trrr, trrrrr, on, on, on, on, wooawooawoooon, gog, magog,* and God only knows what other barbarian words. (Rabelais [1532] 1990:497)

Though his companions wish to preserve some of these words in oil, Pantagruel says that it is not worth saving what is always plentifully at hand. Can we not see in this scenario the phantasm at the origins of radiophonic art, where the word is embalmed and speech immortalized? Only the slit throat, the terminal loss of body, indeed death, permits an eternal return of the voice. This return is situated at the origins of modernism, where the particular characteristics of recorded sound—disembodiment, alienation, repetition, eternalization, temporal malleability, and so forth—simultaneously transform age-old metaphysical and theological paradigms, and offer unheard of formal and practical aesthetic possibilities.

The French playwright Valère Novarina explains the extreme difficulty in reading Rabelais, a difficulty described in terms of a veritable archaeology of the lived, respiratory, musculatory, enunciatory patterns of the French language: "To read him is to change bodies; it is an act of respiratory exchange, it is to breathe within another's body" (1992:75–76). Believing in the interlocutory presence of the lived body, Novarina is a man of the theatre, and he consequently provides a critique of present-day mainstream radio that might serve as a partial guide for our present concerns:

> They work night and day with immense teams and enormous financial means: a cleansing of the body in sound recording, a toilet of the voice, filtering, tapes edited and carefully purified of all laughs, farts, hiccoughs, salivations, respirations, of all the slag that marks the animal, material nature of the words that come from the human body [...]. (1993:100)

He proposes, in its stead, a new use of the voice that harkens back to Antonin Artaud's own transformation of the vocal arts, as manifested in his radiophonic *To Have Done with the Judgment of God* (1948). Certainly apparent in the productions of Novarina's own theatre, there are indeed also radiophonic works that instantiate this recorporealization of the human voice, all the while achieving a disquieting grafting of mechanical, electric, and electronic possibilities onto the strictly human potentials of sound recording and transmission—an artificial transmogrification of respiratory patterns and vocal intonations.

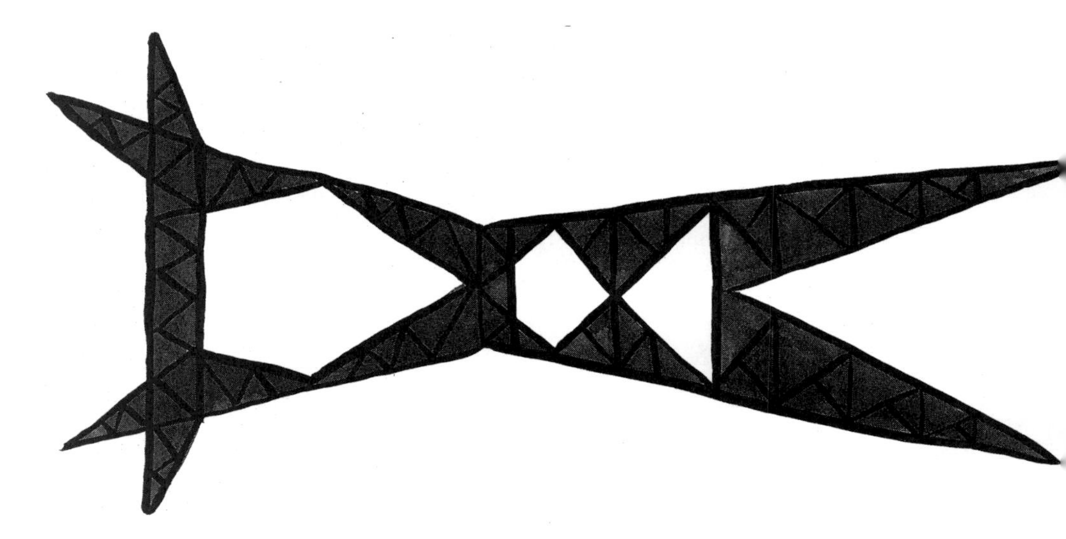

Artwork by Zaven Paré

There is no single entity that constitutes "radio"; rather, there exists a multitude of radios. Radiophony is a heterogeneous domain, on the levels of its apparatus, its practice, its forms, and its utopias. A brief and necessarily incomplete sketch of some possibilities of nonmainstream concepts of radio will give an idea of this diversity: F.T. Marinetti—"wireless imagination" and futurist radio; Velimir Khlebnikov—revolutionary utopia and the fusion of mankind; Leon Trotsky—revolutionary radio; Dziga Vertov—agit-prop and the "Radio-Eye"; Upton Sinclair—telepathy and mental radio; Bertolt Brecht—interactive radio and public communication; Rudolf Arnheim—radiophonic specificity and the critique of visual imagination; the labyrinthine radio narratives of Hörspiel; William Burroughs—cut-ups and the destruction of communication; Glenn Gould—studio perfectionism and "contrapuntal radio"; Marshall McLuhan—the primitive extension of the central nervous system; the diversity of community radios; free radio; guerrilla radio; pirate radio; radical radio...

As such, every "radio" determines an ideal world, though some such domains deal explicitly with the issues of utopia and dystopia, as is evinced in this volume: Richard Foreman's selection from *Hotel Radio* evokes, as does all his theatrical work, the strangeness at the core of the quotidian; Toni Dove's *Casual Workers, Hallucinations, and Appropriate Ghosts* creates an aural evocation of high-tech street erotics; and Lou Mallozzi's *Lingua Franca*, as well as Kaye Mortley's

Around Naxos, offer sonorous investigations of the unique relations between topography, history, language, and experience established by audio montage.

The 19th century was the epoch in which new metaphors of transmission and reception, as well as novel modes of the imagination, were conceived. The "animal magnetism" of Mesmerism was replaced in the 19th century by the spiritualist manipulation of electric waves in the ether, destined to merge with the psychic waves of the departed, such that electricity would permit contact with the afterworld. Walt Whitman, already by 1855, announced "I Sing the Body Electric" as one of the poems in *Leaves of Grass*; Charles Cros would link his lyrical, nostalgic love poetry to his discoveries that would fix time and space: the color photograph and sound recording; Villiers de l'Isle-Adam would reconstitute, in the antitechnological backlash of his *L'Eve future*, a key modernist paradigm following Cartesian mechanistic philosophy, that of the human as machine. Edison would, of course, realize all these fantasies with his invention and successful marketing of sound-recording devices.

At the turn of the 20th century, these new modes of communication, sound production and reproduction were already part of the contemporary psyche: Henry Adams included in his 1907 autobiography, *The Education of Henry Adams*, a chapter entitled "The Virgin and the Dynamo"—nothing better expresses the difference between ancient and modern paradigms of aesthetics and ontology, where the rapidity and excitation of electric power serves as the new symbol of a body now ruled by technology, without divine interference. The virgin/dynamo opposition effectively expresses the different paradigms to be established nearly a quarter of a century later at the interior of radiophonic art. The classic theatre is a stage of history, theology, and metaphysics, of the body given to God and the Virgin, to nature and culture—the body imbued with life-force. The dynamo, to the contrary, is something quite other, creating a new current, flow, circulation, excitation—a force closely allied with the destructive powers of technology. Electricity transformed the very form of the imagination through which we discover our contemporary utopias and dystopias. The first group of essays presented here suggest varied proto-radiophonic phantasms situated at the threshold of this paradigm shift, suggesting a revision of genealogy of the audiophonic arts: In "Erotic Nostalgia and the Inscription of Desire," I attempt to reveal the libidinal and structural relations between the desire to fit forever the eroticized voice and transformations in 19th-century French lyrical poetry; Alexandra Keller's "Shards of Voice" exhibits the perverse, variegated, and intertwining phantasies concerning voices and heads, talking and otherwise; both Mark Roberts's "Wired: Schreber As Machine, Technophobe, and Virtualist" and Christof Migone's "HeadHole" add to a growing literature on the aesthetic implications of psychopathological symptoms and syndromes. These essays help chart the psychological and sociological transformations of the role of the voice within the European symbolic system at the moment of the invention and dissemination of sound recording, changes which imply a radical epistemological shift in the constitution of memory, temporality, and knowledge.

To continue this skeletal history, bringing radiophony into high modernism, the date 2 February 1948 is crucial. This is the moment of the nonevent that remains pivotal in radiophony, the suppression of Antonin Artaud's scheduled radio broadcast of *To Have Done with the Judgment of God*. This year also marks the origin of modern radiophonic and electroacoustic research and creativity, for it was at this moment that magnetic recording tape was perfected and became available for artistic purposes. The confluence of these two events—Artaud's final attempt to void his interiority, to transform psyche and suffering and body into art; and the technical innovation of recording tape, which henceforth permitted the experimental aesthetic simulation and disar-

ticulation of voice as pure exteriority—established major epistemological and aesthetic shifts in the history of art.

Though the radiophonic voice is "disembodied," the body is never totally absent from radio, while it is often radically disfigured, transformed, mutated. The body is neither purely natural nor purely textual, but rather the primal symbolic system that articulates nature and culture. As transformed by the re-recording, looping, and feedback capabilities of sound engineering (especially given the subliminal, microphonic levels of digital sampling), the human voice in radiophonic art (and, by extension, in certain extreme examples of experimental cinema) will project the voice of "nobody," which like Artaud's "body without organs," from his radiophonic *To Have Done with the Judgment of God*, is proposed as an antidote to the ills that beset the fragile, tortured body in pain. We must therefore rethink the radio in terms of a potentially disarticulatory—and no longer articulatory—site of the symbolic, not representing the body but rather transforming or annihilating it. In "Stein's Stein," a piece of theoretical fiction, I detail the serendipitous disjunction between thought and enunciation in an experimental practice that served as a model for the early avantgarde.

Several of the works presented here speak to these issues: Ellen Zweig's "Mendicant Erotics" is a narrative of aleatory relations between erotic encounter and geographic location, suggesting an allegory for constituting a libidinal radio space; John Corbett and Terri Kapsalis's "Aural Sex: The Female Orgasm in Popular Sound" charts out a theory of gender difference in relation to broadcast musical voices, while Mary Louise Hill's "Developing *A Blind Understanding*" offers a parallel analysis, from the point of view of semiotics, dealing with the constitution of gender difference in narrative radio; and both previously mentioned texts by Toni Dove and Alexandra Keller set forth eroticized phantasms aligned with the formal properties of radiophony, while Mark Roberts' analysis of Schreber's psychosis reveals the underlying connections between eroticism and theology, in a delirium where the epochal shift from mechanics to electronics is already seen to be inscribed in the unconscious. Finally, Fred Moten's "Interpolation and Interpellation" examines the psycho-political effects of "engaged" listening as a mode of empowerment.

Certain radical experiments in radiophony, those of concern to us here, suggest the broad potential of radio beyond the various stultifying "laws" that guide mainstream radio: the law of maximal inoffensiveness, the law of maximal indifference, the law of maximal financial return. A sort of perverse specialization—perhaps a manifestation of what Gilles Deleuze in *The Logic of Sense* speaks of as a "logic of the particular"—reigns in certain contemporary pirate radio stations, which determines the margins of aesthetic culture. These experimental possibilities may even operate at the very interior of mainstream, government, military, or commercial radio, rare as they may be: parasites and viruses that determine yet other limits, functions, and pleasures of radiophonic art.

Every new medium first contains and disseminates the forms and content of past media, well before ever revealing its own aesthetic potential. Radio was no exception, and present history has barely changed the situation. In his novel *Les larmes de pierre* (Tears of Stone), Eugène Nicole recounts a charmingly naive phantasm of the radio. The narrator, speaking of his childhood years on the French island of Saint-Pierre during the 1940s, reveals the following:

> Maryse and I now know that the announcer didn't live in the radio. For a long time, at Jacquet's place, we imagined that the radio's interior was arranged like a miniature apartment where, at the same hour each evening, seated on a sofa, after having placed a record on the gramophone, Pointe-Fine spoke to us [...]. We readily admitted that *in the*

radio—like in our dollhouses and our cardboard farms, which always had one wall missing, so that we could serve the children's refreshments, or put animals inside, stuck into the gaps by little wooden pegs—there reigned a different scale of peculiar realities. It was more difficult, however, to understand how that big asparagus Pointe-Fine, with his basque beret and his too-long raincoat, was to be found a half-hour later, not only in our radio, but in all the radios of the city. "The mystery of the Eucharist," exclaimed The Old Woman, raising her eyes to the heavens to underline our ignorance, or to ask pardon of God for this blasphemous parallel, which didn't hinder her from adding, "Like the body of Jesus, while present in each host, is in all the others at the same time." (Nicole 1991:22–25; my translation)

What is at stake is not merely imagination as rememoration, as the reproduction of what already exists, but rather imagination as creative act. Radio is the ideal medium to establish such a poetics and ethics, given its infinite overture to imaginative conjecture and visual discord. Yet seldom is such aesthetic openness manifested or even encouraged in modern media; ironically, mainstream radio uses all of its efforts to deny this poetic source of creativity by restricting radio to old musical and theatrical conventions, by remaining a "clean" medium.

However sophisticated the montage, most works for radio never surpass the conditions of music, theatre, and poetry—radio rarely realizes the potentials specific to the radiophonic apparatus. For radiophony is not only a matter of audiophonic invention, but also of sound diffusion and listener circuits or feedback. Whence the paradox of radio: a universally public transmission is heard in the most private of circumstances; the thematic specificity of each individual broadcast, its imaginary scenario, is heard within an infinitely diverse set of nonspecific situations, different for each listener; despite radio auditors' putative solidarity, they remain atomized, and the imagination is continually reified. The Old Woman is correct: The experience of radio is indeed mystifying, though on a far more mundane level than her analogy would suggest.

In contrast with Eugène Nicole's childhood fantasy, consider the following description by the contemporary radio artist, Gregory Whitehead, from "Radio Art Le Mômo":

> Radio Talking Drum—an utopian transposition that loves to forget. *Most* forgotten are the lethal wires that still heat up from inside out, wires that connect radio with warfare, brain damage, rattles from necropolis. When I turn my radio on, I hear a whole chorus of death rattles: from stone cold, hard fact larynxes frozen at every stage of physical decomposition; from talk show golden throats cut with a scalpel, transected, then taped back together and beamed out across the airwaves; from voices that have been severed from the body for so long that no one can remember who they belong to, or whether they belong to anybody at all; from pop monster giggle-bodies guaranteed to shake yo' booty; from artificial folds sneak-stitched into still-living throats through computer synthesis and digital processing; from mechanical chatter-boxes dead to begin with; from cyberphonic anti-bodies taking flight and crashing to pieces on air. (1991:145)

The man-in-the-radio is countered by the radio-in-the-man. Like Nicole, Whitehead recognizes radio's intimate coupling with sundry nostalgias and forms of death—radio as an electronic *memento mori* for a modern age and a thoughtless public. It is in regard to these new creative possibilities that the core of this *TDR* Reader lies, precisely in those texts that deal specifically

with questions of the ontological and aesthetic specificity of radiophonic montage: Susan Stone's "Cat's Cradle," René Farabet's "From One Head to Another," Joe Milutis's "Radiophonic Ontologies and the Avantgarde," Douglas Kahn's "Three Receivers," Dwight Frizzell and Jay Mandeville's "Inaudible Postscript," and Gregory Whitehead's conversation with Jérôme Noetinger, entitled "Radio Play Is No Place."

This volume was conceived to play a certain role in the current dialog about radio. Considerations of mainstream radio have been for the most part excluded from aesthetic and cultural discourse, and the history of experimental radiophony has until recently been utterly repressed. At this moment that academic and museological recognition is belatedly occurring, we offer the present project as an attempt to complicate such matters. We are concerned with conditions of transmission, circuits, disarticulation, degeneration, metamorphosis, mutation—and not communication, closure, articulation, representation, and simulacra. As the inevitable canonization of the field transpires, we wish to keep its margins fluid. Whence the concern with the occasionally incompatible yet increasingly crucial domains of ontological heterogeneity, disjointed signifiers, broken circuits, dead air, disembodied voices, audio uncanny, linguistic contortions, noise, and spiritualist macabre. The range of these essays—offering but a selection of the vast array of topics currently being explored—should make clear that radio is not merely a communications conduit, but rather a heterogeneous mix of technological progress and aestheticized desire, intermedia mixes and societal restrictions, broadcast possibilities and suppressed histories.

This collection of essays ends on two iconoclastic notes, Brandon LaBelle's analysis of contemporary noise music, "Music to the 'nth' Degree," and G.X. Juppiter-Larsen's "More Facts on the Polywave," for radiophony is the iconoclastic, or at least iconophobic, art form par excellence. Shattered icons create radical schisms which establish new circuits. I might conclude this introduction in an analogous manner, by repeating a joke remembered from my childhood. Both heard on the radio and referring to the radio, the following minimal dialog generated mysteries that incited my earliest reflections on radiophony:

> Two elephants are sitting in a bathtub. One elephant says to the other, "Please pass the soap." The other elephant responds, "No soap, radio."

Nonsense!? I now await, in response, other narratives, other paradoxes, other noises and silences.

Note

1. This introduction is in part derived from two of my earlier texts: "Broken Voices, Lost Bodies" in *Perverse Desire and the Ambiguous Icon* (1994); and "Radio Phantasms, Phantasmic Radio," in *Phantasmic Radio* (1995).

References and Select Bibliography

Allison, David, et al.
1990 "Nonsense." *A+T* [*Art + Text*] 37. Sydney and New York: *A+T* and The Whitney Museum.

Augaitis, Daina, and Dan Lander, eds.
1994 *Radio Rethink: Art, Sound and Transmission.* Banff: Walter Phillips Gallery.

Adriaansens, Alex, et al.
1992 *Book for the Unstable Media.* 's-Hertogenbosch, The Netherlands: V2 Organization.

Davies, Shaun, et al.
1992 *Essays in Sound* 1. Darlinghurst, Australia.
1995 "Technophonia." *Essays in Sound* 2. Darlinghurst, Australia.

de Guerre, Marc, and Janine Marchessault, eds.
1991 "Sound." *Public* 4/5.

Deleuze, Gilles
1990 *The Logic of Sense*. New York: Columbia University Press.

Harrison, Martin, et al.
1989 "Sound." *Art & Text* 31.

Kahn, Douglas, and Gregory Whitehead, eds.
1992 *Wireless Imagination: Sound, Radio and the Avant-Garde*. Cambridge, MA: MIT Press.

Lander, Dan, ed.
1992 "Radio-phonics." *Musicworks* 53.

Lander, Dan, and Micah Lexier, eds.
1990 *Sound By Artists*. Banff: Walter Phillips Gallery.

Mallozzi, Lou, ed.
1994 "Audio Art." *P-Form* 33.

Miller, Toby, ed.
1992 "Radio-Sound." *Continuum* 6, 1.

Mediamatic
1992 "Oor=Era." *Mediamatic* 6, 4.

Nicole, Eugène
1991 *Les larmes de pierre*. Paris: Bourin.

Novarina, Valère
1992 "Chaos." Translated by Allen S. Weiss. *Alea* 2:75–76.
1993 "Letter to the Actors." Translated by Allen S. Weiss. *TDR* 37, 2 (T138):95–104.

Rabelais, François
1993 [1532] *Gargantua and Pantagruel* Book IV. Translated by Burton Raffel. New York: Norton.

Strauss, Neil, and Dave Mandl, eds.
1993 *Radiotext(e)*. Semiotext(e) 16.

Szeemann, Harald, ed.
1976 *Junggesellenmaschinen/ Les machines célibataires*. Venice: Alfieri.

Weiss, Allen S.
1994 *Perverse Desire and the Ambiguous Icon*. Binghamton, NY: SUNY Press.
1995 *Phantasmic Radio*. Durham, NC: Duke University Press.

Whitehead, Gregory
1991 "Radio Art Le Mômo: Gas Leaks, Shock Needles and Death Rattles." *Public* 4/5:140–49.

Artwork by Zaven Paré

Erotic Nostalgia and the Inscription of Desire

Allen S. Weiss

The deux ex machina *took the place of metaphysical comfort.*
—*Friedrich Nietzsche*
The Birth of Tragedy *([1872] 1967:109)*

Sacred love is often transmuted into profane desire, as when Monteverdi surreptitiously transformed Ariadne's lament into that of the Virgin at the foot of the cross (*Lamento d'Arianna*). Towards the end of 1885, Charles Cros and Villiers de l'Isle-Adam together possessed a scruffy fox terrier they named Satan, which they paraded around Paris, claiming that the dog was the receptacle of Baudelaire's soul. Yet given Villiers' technological fantasies and Cros's phonological inventions, this gesture was decidedly anachronistic. Where, today, do we dare place Baudelaire's spirit, or, for that matter, the spirits of those we desire, or love?[1]

The most extreme phantasms often originate in the most extreme resistance, as is often the case in paranoia, and as was the case of the reactionary 19th-century critique of technological progress, with all that this implied for the arts. In 1874, Villiers de l'Isle-Adam wrote one of his *Contes cruels,* entitled "La machine à gloire" [The Glory Machine], dedicated to Stéphane Mallarmé. This diatribe against modernity—motivated in part by a mounting indignation and ressentiment in regard to his theatrical failures—proffers a prototypical manifestation of the theatre of cruelty. Villiers suggests that in the theatre, the *claque,* the hired clappers, constitutes a deception necessary to the success, indeed to the very existence, of the production. The claque is deemed an art form in itself, manifesting the entire gamut of expressivity, such that spectatorial reaction is transformed into art. Beyond the varied types of clapping, there are also a myriad of vocal effects: the initial, basic *bravo,* is soon transformed into *brao;* one then passes on to the paroxysmic *Oua-Ouaou,* which finally evolves into the definitive scream, *Brâ-oua-ouaou,* nearly a bark. But, in fact, these are still only the most basic effects; there exists a full range of special effects of which the claque is capable, including such refinements as:

> Screams of frightened women, choked Sobs, truly communicative Tears, little brusque Laughs [...] Howls, Chokings, Encore!, Recalls, silent Tears, Threats, Recalls with additional Howls, Pounding of approbation, uttered Opinions, Wreaths, Principles, Convictions, moral Tendencies,

> epileptic Attacks, Childbirth, Insults, Suicides, Noises of discussions (Art-for-art's-sake, Form and Idea), etc. ([1874] 1983:108)

The final word of this art is when the claque itself shouts, "Down with the claque!" and then applauds the piece as if they were the real public. As Villiers explains, "The *claque* is to dramatic glory what the Mourners are to Suffering" (108).

Even so, this is but mere art; the aleatory effects of the claque can, in fact, be eliminated, according to Villiers, by mechanizing the process. This is the "Glory Machine," which will be constituted by the auditorium itself, where the entire audience will surreptitiously be transformed into the claque. In this apparatus, the sound effects are perfected by multiplying the presence of gilded angels and caryatids, whose mouths bear phonographic speakers to emit the appropriate sounds at critical moments; the pipes that supply the lamps with gas are augmented by others to introduce laughing gas and tear gas into the auditorium; the balconies are equipped with mechanisms to hurl bouquets and wreaths onstage; spring-operated canes are hidden in the feet of the chairs, so as to reinforce the ovations with their striking. In fact, the apparatus is so powerful that it can, literally, bring down the house, such that the theatre would be totally destroyed!

In this 19th-century aesthetic dystopia, where art is sublated into industry, Villiers manages to eradicate the need for actor, scenario, and scene. The spectacle is reduced to audience reaction, in what is not quite a conceptual theatre, but rather a purely sensual stagecraft. This ironic, unwittingly modernist event creates the immediate yet ephemeral inscription of sensation directly on the spectator's body, not unlike the psychedelic "inner" theatres of 1960s drug culture—an iconoclastic technique of theatreless theatre that effects a counter-memory, counter-spectacle, and counter-symbolic.

This technique is coherent with physiological experimentation and theorization of the 19th century, which understood perception to be possible in a nonreferential manner. Such was demonstrated by experiments proving that impressions of light may be produced without any visual stimuli whatsoever, either by mechanical, electrical, or chemical means.[2] To seek the aesthetic limits of such techniques would be to theorize not the sublime but the countersublime, where temporality is constituted by reflexively closed-in-upon physiological rhythms and thresholds; where consciousness, subsumed by pure presence, eschews all transcendence; where the imagination exists in direct proportion to somatization; and where, purged of language, the symbolic code is abolished. Narration is obliterated, time nullified, and the psychic mechanism thrust into a solipsism rivaling that of the mystics, inaugurating the oxymoron of an innate apocalyptic sublime. In what would appear to be an ultimate extrapolation of Baudelaire's utopia of an "artificial paradise," the Romantic sensibility merges with a nascent scientific positivism to indicate a major trajectory of modernist performance.

In the year 1801, Pierre Giraud, seeking methods to ameliorate the horrendous conditions of the Parisian cemeteries, wrote *Les Tombeaux, ou essai sur les sépultures*:

> In which the author recalls the customs of ancient peoples, mentions briefly those observed by the moderns, and describes procedures for dissolving flesh and calcining human bones and converting them into an indestructible substance with which to make portrait medallions of individuals. (Ariès [1977] 1981:513)

In this project, Giraud cites the work of a 17th-century inventor, Becker, in whose *Physica subterranea* we find the notion of transforming mummies of fat into mummies of glass—what will be described as, "The Art of Vitrifying Bones." Saved from the horrors of the tomb, the beloved will remain forever—in form and substance—by means of this new innovation in the cult of the dead. Philippe Ariès explains that this project confuses the language of two different periods and two distinct paradigms of treating the dead: "the period where the cadaver promised to reveal to anyone who dissected it the secrets of life and the period when the cadaver gave to anyone who contemplated it the illusion of a presence" ([1977] 1981:516). Through which of these models can we most effectively mediate the death of others, as well as our own, unrepresentable, death? All necrologies serve to nurture the memories of the departed, as well as to prefigure signs of the memories that we are to become—yet some are decidedly more Romantic, and more *romantic*, than others.

The 19th century would mark a great paradigm shift in our relations with the dead, where the eternal desire to maintain contact with the dead would finally, thanks to photography and sound recording, offer means of resolution that were simultaneously indexical, iconic and symbolic. The first book of Charles Cros' collection of poetry, *Le Collier de griffes* [The Necklace of Claws], is entitled *Visions*, of which the introductory poem, "Inscription," simultaneously describes Cros' scientific discoveries and his erotic nostalgia:

> J'ai voulu que les tons, la grâce,
> Tout ce que reflète une glace,
> L'ivresse d'un bal d'opéra,
> Les soirs de rubis, l'ombre verte
> Se fixent sur la plaque inerte.
> Je l'ai voulu, cela sera.
>
> Comme les traits dans les camées
> J'ai voulu que les voix aimées
> Soient un bien, qu'on garde à jamais,
> Et puissent répéter le rêve
> Musical de l'heure trop brève;
> Le temps veut fuir, je le soumets. ([1908] 1972a:25–26)
>
> [I wanted the tones, the grace,
> Everything reflected in a mirror,
> The drunkenness of an opera ball,
> Ruby evenings and green shadow
> To be fixed on the inert plate.
> I wished it, so shall it be.
>
> Like the features on a cameo
> I wanted the beloved voice
> To remain a keepsake, forever cherished,
> Repeating the musical
> Dream of an hour all too brief;
> Time wishes to flee, I master it.]

Cros experimented with two techniques to stop and fix time—color photography and sound recording—having conceived of the "paleograph," a sound-recording device, in 1877, the same year that Edison inventioned the phonograph. This machine established the possibility of eternally fixing and reproducing the sonorous spectacle. Thus, in antithesis to the radical ephemera of Villiers's glory machine, the paleograph would inscribe a past become infinitely representable and malleable. Immortality would be achieved at the cost of disassociation, decomposition and decorporealization—beyond any possible resurrection of the body.

No longer, as in his early poem, "La dame en pierre" (The Woman in Stone) would Cros's amorous nostalgia need suffer the stultifying, melancholic effects of petrification, as was typical of attitudes toward death in prerecording epochs, such as those of Giraud and Becker:

> La mort n'a pas atteint le beau.
> La chair perverse est tuée,
> Mais la forme est, sur un tombeau,
> Perpétuée. ([1873] 1972b:41)

> Death hasn't touched beauty.
> The perverse flesh is killed,
> Yet the form, upon a tomb,
> Is perpetuated.

Henceforth, the serendipitous modalities of a lover's discourse shall no longer be bounded by mere nostalgia in the face of death. A new, radical dissociation of form and content now intervenes, such that even if passion cannot conquer time, it can, however, avail itself of a particularly simulacral relic: the eternally perpetuated voice of the beloved.

Such phantasies had their metaphysical correlates. In 1881, Nietzsche—seeking the atmospheric electricity that he hoped would be a decisive factor in curing his varied ills—traveled to Sils-Maria, where he suffered the intuition of the Eternal Return. It received its major expression in *Thus Spoke Zarathustra*:

> That time does not run backwards, that is his wrath. Revenge is the will's ill will against time and its "it was." "It was"—that is the name of the will's gnashing of teeth and most secret melancholy. The will cannot will backwards; and that he cannot break time and time's covetousness, that is the will's loneliest melancholy. To redeem those who lived in the past and to recreate all "it was" into a "thus I willed it"—that alone should I call redemption. All "it was" is a fragment, a riddle, a dreadful accident—until the creative will says to it, "But thus I willed it." ([1883–85] 1980:249–54)[3]

Thus spoke Zarathustra. Thus wrote Nietzsche. It is not by chance that the first major modern European contestation of linear temporality (other than Schopenhauer's metaphysical orientalism) was contemporaneous with the invention of recording technologies. The elimination of temporality is a manifestation of the revenge of a strong poetic will, a reaction against time itself. Poetic substitution (replacement by tropes) is the transformation of the "it was" into an "it is," with the subsequent transmogrification of this "it is" into an act of volition. Within this context, which implies a shift in both the classic rhetorical and ontological orders, the figure of *hysteron proteron* emblematizes a reversal of Western metaphysics, heretofore ruled by the ancient dreams of temporal reversal and time travel. Of all the arts, it is precisely those based upon recording technologies, permitting a radical plasticity of time, that most vividly meet these paradoxical conditions of renewal and creativity, reversal and transmutation.

While Villiers's glory machine offered the minimal aesthetic model of an imageless, indeed iconoclastic, realm of pure affect, the 19th century valorized its antithesis: the totalizing presumptions and effects of Wagnerian opera, variously celebrated in the circles of Baudelaire, Nietzsche and Mallarmé.[4] The architectural constitution of the "mystical abyss" (the site where the orchestra is dissimulated) separates spectator from proscenium and real from ideal, creating the conditions whereby a distant dream-vision arises. The mythical Wagnerian phantasmagoria was made possible by a major technological innovation: the electric light, permitting a myriad of effects, notably that of totally darkening the auditorium and lighting the stage with great precision.[5] Theodor Adorno, in a passage concerning *Tannhäuser*, illustrates the intimate relations between Wagnerian aesthetics and technology:

> The standing-still of time and the complete occultation of nature by means of phantasmagoria are thus brought together in the memory of a pristine age where time is guaranteed only by the stars. Time is the all-important element of production that phantasmagoria, the mirage of eternity, obscures. ([1952] 1985:87)

Wagner desired the aesthetic paradox of eternalization through the ephemeral, not unlike the ontogenetic manifestation and perpetuation of myths within the dream-work. On a decidedly less mythic (though equally apocalyptic and megalomaniacal) level, he wished that *The Ring* be performed but three times, and that afterwards the libretto, scenery, and even the theatre itself be destroyed by fire—a veritable glory machine, in tune with fin-de-siècle apocalyptic imagination.

The opposition between the purely imageless, iconophobic, physical intoxication of Villiers' glory machine with the dreamlike, imagistic phantasmagoria of Wagnerian opera delineates what shall become a major modernist aesthetic paradigm: Nietzsche's distinction between the Dionysian and the Apollonian. The Apollonian is the world of pure form and dreams; to the contrary, the Dionysian exists emotively, through intoxication, without images. In the latter—as in the purely corporeal effects experienced by the audience within the glory machine it is the case that:

> the entire symbolism of the body is called into play, not the mere symbolism of the lips, face and speech but the whole pantomime of dancing, forcing every member into rhythmic movement. Then the other symbolic powers suddenly press forward, particularly those of music, in rhythmics, dynamics, harmony. ([1872] 1967:40)

It is precisely the function of drama and opera to transfigure Dionysian intoxication into Apollonian vision, to transform libido into sign. Dionysus is the body marked by difference, disorder, disintegration, forgetting; Apollo is the body traced by identity, order, the Gestalt of good form, and memory. Though antithetical, these gods are intimately linked; indeed, Apollo is but a manifestation of Dionysus. Thus, to pay tribute to one God while ignoring the other is to court disaster; psychically, it is to instill repression, or even the foreclosure of madness (see Weiss 1989:3–11).

Mallarmé—representing the inner limit of that great surge of musication in French poetry of the late 19th century, extending from Verlaine and Rimbaud through Valéry—well understood the exigencies of the relation between sound and image in Wagner. In his celebratory text, "Richard Wagner—Rêverie d'un poëte français," Mallarmé writes of the sublime, totally generative aspect of Wagner's music: "an audience would have the feeling that, if the orchestra were to cease exercising its control, the mime would immediately become a statue" ([1885] 1945a:543). This inversion of the myth of Galatea is telling. Rameau's *Pygmalion* offers the scenarization of an ontological category error transformed, through wish fulfillment, into aesthetic delight. Here, passion is projected as beauty, in the form of a statue animated by the artist's desire. And this desire is choreographed: the statue of Galatea takes her very first steps to the sound of music, as the three Graces teach her to dance, before she even learns to walk. As Philippe Beaussant explains: "In the sublime scene where the statue of Galatea is animated and comes to life, Rameau's work seems to assume the totality of its signification, which is precisely the intrusion of the sculptural upon music and dance upon song" (1988:136). We may extrapolate, and argue that no art exists without the supplement of another (or of perhaps of all the other) arts. Furthermore, those arts that constitute *scenarization* (such as architecture, with its sonic and visual formalizations and delimitations) are *a fortiori* present as the tacit precondition of the performing arts.

Mallarmé's work is imbued not only with the musical metaphor, but indeed with one of the most subtle and precise senses of musicality in modern French poetry. Yet Mallarmé, arch Apollonian, had no such need of music to animate his verse: the musication or musicality of his poetry sufficed. However, this is a musicality radically divorced from expression. Though he claims "every soul is a melody, which must be renewed" ([1895] 1945b:363) and "every soul is a rhythmic knot," ([1894] 1945c:644) considerations of the soul were in fact anathema for Mallarmé. Rather, he sought a poetics where, as was evidenced in the preceding chapter, "the pure work implies the elocutionary disappearance of the poet, who cedes the initiative to words [...] replacing the perceptible respiration of the ancient lyrical breath or the personal enthusiastic direction of the sentence" ([1895] 1945b:366). Unlike that Nietzschean "blissful ecstasy" which results from the Dionysian collapse of the *principium individuationis* ([1872] 1967:36), the Mallarméan disappearance of the author occurs by virtue of the absorption of lyrical voice within the text, with the consequent loss of lived voice and the exteriorization of text as object. Unlike Cros—whose lyricism was but the shadow of a dreamt reconciliation between voice and image, body and memory, yet for whom the stone figure of the beloved could never be reincarnated—Mallarmé would not valorize the disincarnate voice.

On 11 September 1898 Stéphane Mallarmé was interred at the cemetery of Samoreau. In the presence of such poets and friends of the deceased as Henri de Régnier, Catulle Mendès, and José Maria de Heredia, it was Paul Valéry who gave the funeral address, in the name of the French poets of his generation. As he stepped forth to speak, the words stuck in his throat, he was unable to articulate, and he remained speechless. Mallarmé died of a "spasm of the larynx," and the orator of his obsequies remained mute, choking on his words! An appropriate symptom, especially if one considers that Valéry was later to explain that poetry, "is first born in the muscular articulations of the throat, which in fact finds itself in knots because poetry is dead" (in Michel 1984:51).[6] This hyperbolic symptom bearing witness to the anxiety of influence is well explained in a letter Valéry wrote years later, where, speaking of Mallarmé, he exclaims, paraphrasing a notorious remark of the emperor Caligula recounted by Suetonius in *The Twelve Caesars*: "I adored that extraordinary man at the very same time that I found his to be the one head—priceless—to be chopped off in order to decapitate all of Rome" (1973:60). This morbid reaction is all the more telling as Valéry finds in Mallarmé the ultimate limit of the poetic art, a limit that Valéry could surpass only in seeking an impossible linguistic perfectibility. As he wrote in his diary: "I find that poetry interests me only as the research of a very small problem whose solution is most improbable: syntax × musique × conventions. As for the rest—the imagination, physics, and mathematics are far more exciting and rich, etc." (1973:171).

Such richness is at the very core of Valéry's epistemology, one not without relation to that of Nietzsche: "Everybody has a metaphysics (always much stranger than the avowable metaphysics). This metaphysics depends enormously on the *unity* of the measure unconsciously adopted by each person" (1973:69). But this is a false unity, insofar as there is never any single "unity of measure," nor any metaphysical absolute, in things poetic, linguistic, existential. Whence Valéry's avowal that, "I speak a thousand languages. One for my wife, one for my children, one for the cook, one for my ideal reader—and each category of friends, merchants, businessmen..., his own" (1973:395). This reduction to solipsism through an individualistic metaphysics entails the following tautology: "I do not create a 'System'—My system is me" (1973:208). This must be contrasted with Mallarmé's quasi-Hegelian "system," which is but the *simulacrum* of a metaphysics, a framework for organizing the aleas of existence within language. Worthy of Igitur's desolation, Mallarmé's "system" is ultimately inseparable from the contingencies of his psychological crisis, as the poem is inseparable from chance. Thus the core of Valéry's theory of poetry was an ironic inversion of Mallarmé. While Mallarmé transformed a psychic crisis—with all of the ramifications of its unconscious machinations—into the "system" that would presumably organize his poetry, Valéry would constitute his poetic work as a subcategory of a more general intelligence, inspired by cognitive, indeed scientific, paradigms. Whence two modes of Apollonianism, the former based on the repression and sublation of Dionysis, the latter on the sheer disavowal of the poetic efficacies of this drunken and unkempt god.

One might imagine that for Valéry, the poetry of Villiers's glory machine would not be manifested within the frenzied spectatorial reactions elicited by this insidious theatre (as would be the case if one were to conceive of this invention in terms of Artaud's theatre of cruelty, for example), but rather in terms of the cold poetics of its architectural and technological construction. While Valéry's ideal was stated as, "Poetry—undulatory mechanics!" (1974:1111) the reality of poetry is conceived in quite different terms. Valéry—who defined poetry as, "that prolonged hesitation between sound and sense" (1974:1065)—insists that, "a poem or an extraordinary idea are accidents in the current of words" (1974:987). In a rather Nietzschean turn, he claims further that, "The book, writing, is for me an *accident*—the artificial limit of a mental development" (1973:269); and, "Words do not hide mysteries, but awkwardness, incoherence, chance" (1973:390). Yet this is not to confuse poetry and passion, as Valéry maintains the Apollonian/Dionysian distinction:

> Poetry is the attempt to represent, by means of articulated language, those *things* or that *thing* that cries, tears, silences, caresses, kisses, sighs, etc., obscurely attempt to express, and that objects, insofar as they have the appearance of life, or of a presumed design, seem to want to express. This thing is not otherwise definable. It is of the nature of energy—of excitation, that is to say of *expenditure*. (1974:1099–1100)

For Valéry, poetry exists as a modality of language, as a *representation* of the underlying emotions, a fact made evident by a marginal note attached to the word "poetry" in the above citation, qualifying it as: "deep tenderness, exquisite cold, no grief, no tears." The limits of poetry are the limits of language. The emotions themselves are of a quite different epistemological order. Like Mallarmé, Valéry—in search of the ultimate purity of the poetic-linguistic signifier—wished poetry to be voided of all that is extraneous to language. "That is why I always dreamt of a 'pure literature,' that is to say one founded upon a minimum of direct excitations upon the person and the maximal recourse to the properties intrinsic to *language*. Acute Apollonianism" (1973:272).

Language is no longer seen by Valéry either in terms of a perfectible unity (Classicism) or a malleable expressibility (Romanticism), but rather as a diacritical, differential system, contemporaneous with that of Saussure's structural linguistics: "Language is a statistical ensemble" (1973:416).[7] This realization serves multiple functions: it relativizes the linguistic existence of the individual, inaugurating for each speaking being a unique metaphysics (heralding the death of philosophy); it confines poetry within a calculable realm (delimiting the powers of *poesis*); it thematizes Mallarmé's intuitions concerning the limits of poetry (setting the stage for the possibility of free verse in French); and it absolutizes the role of the aleatory within language (recognizing chance as the very structure of language, thus overcoming Mallarmé's metaphysical anguish—and Valéry's anxiety of influence—by assuming, generalizing, systematizing, and taming chance).

It is precisely in terms of the role of chance that Valéry both wishes to transform (minimize) poetry, as well as overcome his anxiety of influence vis-à-vis Mallarmé. We find, in fact, amidst Valéry's notes on mathematics, a formula that reveals the profound implications of this notion: "One can deal with probabilities without pronouncing the word *Chance*. As one can deal with electricity without pronouncing the word *Frog*" (1974:824). Or, we might add in apparent bad faith: as one can deal with poetry without mentioning Mallarmé. Rather than utilizing the poem to organize or sublate the contingent and aleatory features of existence, as did Mallarmé, Valéry's scientism effectively posits the virtual disappearance of poetry in a sort of general field theory of cognitive activity.

For Valéry, "The delicate point of poetry is the procurement of the voice. The voice defines pure poetry" (1974:1077). This is, of course, the "voice of the poem," not that of the poet, for he equally insists that, "The systematic elimination of what is *voice* [parole] is the capital point of my philosophy" (1973:395). The poetic voice, in its supreme though all-too-rare instances, is pure and imaginary, a "putative" poetic enunciation; the real human voice is fraught with precisely those haphazard qualities and imperfections that are to be eliminated through poetic practice. In an updated "muse theory" of creativity, it would seem that Valéry desired a sort of aural version of Galatea, given his claim that, "The most beautiful poetry bears the voice of an ideal woman, Mademoiselle Soul" (1974:1076). The problem is that lyrical poetry loses its lyricism when not recited. Valéry's narrow and somewhat traditional symbol of poetic inspiration proffers a melodic "rhythmic knot" that is now gendered, following the lineaments of Valéry's desire. From the Romantic nightmare of Poe's "The Oval Portrait" to the Symbolist tragedy of Hadaly, Villiers's "Eve of the future," the perfect simulacrum of the beloved would remain both the allegory of art and the sign of death incarnate. Is it the imagined voice, or rather its reproduction through recording, that shall be the vehicle of erotic nostalgia?

According to Roland Barthes, the "grain of the voice" is constituted by "the materiality of the body speaking its maternal tongue" ([1972] 1982a:238). Whence the voice as articulation of an erotic relation:

> [T]here is no human voice in the world that is not the object of desire—or of repulsion: there is no neutral voice—and if occasionally this neutrality or blankness of the voice occurs, it constitutes a great terror, as if we were to fearfully discover a petrified world, where desire would be dead." ([1977] 1982b:247)

Might we presume that for Barthes, a poem written but unrecited—Mallarmé's dismissal of "the ancient lyrical breath"—would be but a dead letter? But voice is productive, not merely reproductive. According to the psychoanalyst Denis Vasse, "The voice is neither of the order of representation (knowledge) nor of the order of self-presence (place)" (1974:185).[8] Voice articulates body and language, place and knowledge, self and other, the imaginary and the symbolic, by founding an existential limit perpetually transgressed through speech. This transgression can well be imaged, precisely through the prosodic aspects of speech: the sonorous textures of the vocal process evoke a body, a sex, a desire, a death. This transgression—literally taking the form of what Anthony Burgess wrote of as "a mouthful of air"—constitutes nothing other than the multifarious, heterogeneous, and often contradictory processes of consciousness itself.

Though verse is fashioned by voice, there is a distinct futility in Valéry's claim that, "If we better understood this true relation we would know what Racine's voice was like" (1974:1094). This mode of reconstituting or recapturing the lost voice of the dead is an ancient literary and religious device. Witness, for example, the narrator of Proust's *Du côté de chez Swann* recount how his mother would read him to sleep by reciting the prose of Georges Sand:

> [C]areful to banish all smallness from her voice, all affectation that could have hindered the powerful stream of words from being received, she furnished all the natural tenderness, all the ample softness the sentences demanded of those words, sentences which seem to have been written for her voice and which, as it were, remained whole within the register of her sensibility. In order to attack them in the appropriate tone, he found that cordial accent that preexisted and dictated them, but which was not at all indicated by the words. Thanks to this she dampened, as she went along, all roughness in the tense of verbs; she gave the imperfect and the past historic the softness that exists in goodness and the melancholy that exists in tenderness; she directed the sentence that was ending toward the one that was about to begin, sometimes in a hurry, and at other times slowing down the stride of the syllables so as to permit them to enter with a uniform rhythm, even though their quantities differed. She breathed into this so very common prose a sort of sentimental and continuous life. ([1913] 1954:56)

The narrator's mother found—as a theatre director or reciter of poetry would say—the voice of the text, if not of the author. To take yet another example—one more germane to technological rather than human capabilities—is it any more likely that Valéry could reconstitute Racine's voice through phonological analysis, than that professor in Salomo Friedlaender's tale, "Goethe Speaks into the Gramophone," could capture Goethe's voice by digging up the poet's skeleton, reconstructing the larynx, and wiring it to a microphone in order to recapture those vocal vibrations which, though weakened by time, could not have totally disappeared?[9]

Valéry confuses desire with its object, as is evident in his wish to create a "besoin-phénix" (phoenix-impulse), where memory would maintain a constant renaissance of desire, a procurement of the other: (the more I have you, the more I want you) (1974:1136). Such might be the lyrics of a hit romantic show tune, yet they also reveal a darker, unregenerative side of Eros, as Pierre Saint-Amand, in another context, so eloquently explains:

> This imaginary incorporation of the other into the self is invasive, even fatal. It can lead to death, which is at once deliverance, exorcism, and out-fascination. Death is the outcome of desire exhausted, and the outermost limit of confrontation with the other-obstacle. Such is the madness of seduction. (1994:13)

Seduction as unregenerative incorporation exists in chiasmatic intertwining with that poetization of nostalgia common to the work of mourning. Whence the ineluctable relation between Eros and Thanatos. It is precisely because of the mimetic factor in Eros that the phantasmatic origins of recording are so closely linked with morbid amorous nostalgia. For as we know, one of Edison's primary motivations for the invention of sound recording technology was, "for the purpose of preserving the sayings, the voices, and *the last words* of the dying member of the family" (in Harvith and Harvith 1987:1). Words which could now echo forever in both heart and ear.

The linguistic, poetic, and rhetorical effects of the very first sound recording—Edison reciting "Mary had a little lamb," with all its ontotheological connotations—transformed both poetical and metaphysical categories. The effects of amplification, repetition, reversal, dubbing, projection, broadcast, dis-association, and disembodiment quickly equaled those of the most profound theological fantasies. Indeed, if rhetoric hadn't been subsumed by poetics within modernism, the entire category of tropes would need to be rearranged due to the exigencies of recording, with privilege given to many long-forgotten figures. Such a major rhetorical shift would entail a new auditory ontology. In general, the phantasmatic disarticulation and decay of the body is now established according to the accumulation, combination, permutation, and substitution of linguistic elements. All linguistic "aberrations"—especially glossolalia, dissonance, cacophony, invention of pseudo-languages, expansion of vocal timbre—are inflected or "infected" by recording techniques. Thus traditional rhythmic patterns of locution, modulated by corporeal processes (breathing, heartbeat, blood circulation, nervous-system humming) as well as by interruptions of locution (coughing, sneezing, wheezing, gagging, hiccups, borborygmi) are all subject to transformation, highlighting, or suppression by means of the cutting knife of tape montage.[10]

Sound recording inaugurated a new dimension to all possible necrophilia and necrotopias, resuscitating the rhetorical figure of *prosopopoeia*, which manifests the hallucinatory, paranoid, supernatural, or schizophrenic presence of invisible, deceased, ghoulish, demonic, or divine others. These *disembodies* demand a new phantasmatic topography, one which will find its theorization in Gaston Bachelard's *The Poetics of Space*, where he celebrates the topophilia of "felicitous space." Yet he all the while recognizes the disquieting existence of its antithesis, what he terms "oneirically incomplete" dwellings ([1958] 1969:26).[11] Here, we enter the realm of topophobia, of the architectural counter-sublime, the corporeal correlate of which would be the oneirically incomplete body: a condition manifested in the *diasparagmos* of the gods, the body-without-organs, the dismemberments regulated by schizophrenic crises, the sado-masochistic extremes of erotic fantasy, and the acousmetric condition of the impossible radiophonic body.

The antithetical yet complimentary limits of these unrepresentable architectural dystopias mark the limits of modernism: from its anti-Enlightenment inauguration in the secret closed chambers of the Sadian chateaux, to its ultimate diffusion in the vast cosmic expanses of radiophony. These are places of forgetting, of counter-memory: Sade's secret chambers are the sites of invisible orgies and unimaginable tortures, beyond the scope of narrative visuality, offering evidence only through the terrifying sounds emitted; radio is a vast necropolis where the voices of people, both living and long dead, continue to circulate, all the while disintegrating and mixing with each other, in a promiscuous auditory montage.[12] Unlike those delicately orchestrated libertine spaces, the *maisons de plaisance* epitomized by Jean-François Bastide's *La petite maison* ([1758] 1995),[13] Sade's chateaux, and more particularly their inner chambers, constitute the scandalous extreme of the most radical liaisons, the most convoluted combinatory mechanisms. The unexpressed activities that take place there are precisely that textual supplement which would totalize the erotic combinatory system, if such totalization were possible. These closed chambers are the interiorized inversion of the terrifying exterior expanses of radio. Spaces obscene, because haunted by death; sites fascinating, because ruled by pure metamorphosis, juxtaposition, and combination; scenes of excess, because they necessarily extend beyond the limits of any single imagination; realms of seduction, because they permit that phantasmatic projection which is the very ground of mimetic spectatorship; theatres of pornography because of an unspeakable promiscuity; domains of transgression, because symbolic articulation is no longer possible[15] (see Whitehead 1992:253–63; and Hénaff 1978:88–93).

During the 1960s there existed numerous second- and third-run cinemas in Paris, specializing in monster and horror films. In the wings, one could witness, or even participate in, provocative scenes of intense, most often anonymous, erotic activity. Here—as in much 19th-century Parisian bourgeois theatre, though with much less inhibition—the spectators became the spectacle, and the eroticized body became the scene. Several of these sites—such as Le Brady at Château d'Eau and Le Mexico near Clichy—offered a peculiar architectural feature, insofar as the bathrooms (where the private scenarios usually culminated) were located behind the movie screen. Thus, within these scatological *maisons ouvertes* (to ironically coin a phrase), the caresses and couplings of rapid love were dubbed with the inarticulate, inhuman, and disembodied screams of monsters and mutants, vampires and ghouls.

Can we not see in such erotic scenarios an example of the rare confluence of antithetical oneiric spaces, where the intimacy of the closed (albeit public) chamber and the acousmetric presence of the distant, disembodied recorded voice combine to create an oneirically *overdetermined* architecture? Such is a site where both detached Apollonian spectatorship *and* participatory Dionysian drunkenness coexist and coalesce. At the end of the 18th century, the sublime was corporealized through libertinage, demonized by the Terror, and finally interiorized by Romanticism. Now, during the cataclysm of AIDS, the ideals and pragmatics of Eros differ vastly (see Crimp 1989:3–18). I offer these thoughts in memoriam for friends lost; with nostalgia for an eroticism transformed; and as a lament for a terrible new appearance of Thanatos. Given these epochal shifts, what, today, can be the difference between the sublime and the uncanny?[13]

Notes

1. An earlier version of this text appeared in *Essays in Sound #2* (Weiss, 1995a). All translations from the French are my own, unless otherwise indicated.
2. See Jonathan Crary, *Techniques of the Observer* (1992:89–92), for a detailed history of such phenomena.
3. (The present citation is a condensation of Nietzsche's text.) The notion of the Eternal Return was first expressed in 1882, in *The Gay Science*, and its major statement consists of its four enunciations in *Thus Spake Zarathustra*. It should be noted that Zarathustra's inability to enunciate the Eternal Return (it is only alluded to, expressed in dreams, hallucinations, whispers, circumlocutions, ellipses, and pregnant silences, but is never actually stated) points to an ontological double bind at the core of Nietzschean philosophy (see Weiss 1989:18–21).
4. In 1869, Cros dedicated the journal publication of his early poem, "L'orgue": "A Richard Wagner, musicien allemand."
5. On theatre lighting in the 19th century, see Carolyn Marvin, *When Old Technologies Were New* (1988:152–90); Wolfgang Schivelbusch, *Disenchanted Night* ([1983] 1988:203–

21); and Beat Wyss, "*Ragnarök* of Illusion: Richard Wagner's 'Mystical Abyss' at Bayreuth" (1990:57–78). Note that the electric light had from its inception a particularly theatrical destiny, as the first practical electric arc lamps were used in the Paris Opera in 1836.

6. It should be noted that Valéry writes of his poetic influences—characterized as "défenses désespérées"—as being primarily limited to Poe, Rimbaud, and Mallarmé, and as being founded upon his extreme reaction to their works during a period of intense personal crisis during the years 1892 to 1894; (see Valéry 1973:178).
7. In 1943, Valéry wrote: "*A calculated use of chance.*—It is doubtlessly only within literature that this is conceivable" (1974:1139). Though it would take another decade for the music of Cage, Boulez, Stockhausen, and Xenakis to "control chance," it should be remembered that Duchamp was already at work in this domain by the landmark year of 1913, with the creation of his first Readymades and the composition of his *Musical Erratum.*
8. For a particularly acute study of the relation between the body and the imaginary, see Sami-Ali, *Le Corps, l'Espace, et le Temps* (1990).
9. This 1916 tale is cited in Friedrich Kittler, *Discourse Networks: 1800/1900* (1990:230–31).
10. On this modernist rhetorical shift, taking the example of Gregory Whitehead's radiophonic works, see Allen S. Weiss, *Phantasmic Radio* (1995b:60–69, 87–89).
11. Such constructions are perhaps best instantiated by Frederick Kiesler's projects, notably the 1959 model for the *Endless House,* as analyzed in Lisa Phillips, *Frederick Kiesler* (1989).
12. The secret chambers must be distinguished from the *salon d'assemblée* in the Chateau de Silling of *The 120 Days of Sodom,* insofar as the latter constitutes a more classic theatrical space, though one where the audience of libertines, inflamed by the narratrices' tales, soon become actors as they act out their passions. See Anthony Vidler, "Asylums of Libertinage," in *The Writing of the Walls* (1987:103–09).
13. It was suggested to me that the relations between technology and poetics briefly sketched out in this text might appear to be too teleologically oriented. I would answer by evoking Maurice Merleau-Ponty's claim that there are inevitably dead-ends in the historical (and, by extension one might add, in the art-historical) dialectic. I am well aware that such a response would, of course, call into question both the efficacy of Mallarmé's "Hegelianism" as a possible model for any poetics other than his own, as well as censuring the reductiveness and occasional absolutism of a certain tradition of dialectically oriented art and literary criticism. To situate the aesthetic ideal with which this essay concludes in the toilets of a seedy, third-rate Parisian movie theatre would indeed seem to suggest such an impasse, where dialectic dissipates into excess. But hasn't the avantgarde always been precisely what hovers about, or creates, such felicitous spaces?

References

Adorno, Theodor Adorno

1985 [1952] *In Search of Wagner.* Translated by Rodney Livingston. London: Verso.

Ariès, Philippe

1981 [1977] *The Hour of Our Death.* Translated by Helen Weaver. New York: Alfred A. Knopf.

Bachelard, Gaston

1969 [1958] *The Poetics of Space.* Translated by Maria Jolas. New York: Beacon Press.

Barthes, Roland

1982a [1972] "Le grain de la voix." In *L'obvie et l'obtus.* Paris: Le Seuil.

1982b [1977] "La musique, la voix, la langue." In *L'obvie et l'obtus.* Paris: Le Seuil.

Bastide, Jean-François

1995 [1758] *The Little House: An Architectural Seduction.* Translated by Rodolphe El-Khoury. New York: Princeton Architectural Press.

Beaussant, Philippe

1988 *Vous avez dit "baroque"?* Arles: Actes Sud.

Crary, Jonathan

1992 *Techniques of the Observer.* Cambridge, MA: MIT Press.

Crimp, Douglas
1989 "Mourning and Militancy." *October* 51:3–18.

Cros, Charles
1972a [1908] "Inscription." In *Le Collier de griffes*. Paris: Gallimard/Poésie.
1972b [1873] "La dame en pierre." In *Le Coffret de santal*. Paris: Gallimard/Poésie.

Harvith, John, and Susan Edwards Harvith, eds.
1987 *Edison, Musicians, and the Phonograph*. Westport, CT: Greenwood Press.

Hénaff, Marcel
1978 *Sade: L'invention du corps libertin*. Paris: Presses Universitaires de France.

Kittler, Friedrich
1990 *Discourse Networks: 1800/1900*. Translated by Michael Metteer with Chris Cullens. Stanford: Stanford University Press.

Mallarmé, Stéphane
1945a [1885] "Richard Wagner—Rêverie d'un poëte français." In *Oeuvres complètes*. Paris: Gallimard/Pléiade.
1945b [1895] "Variations sur un sujet." In *Oeuvres complètes*. Paris: Gallimard/La Pléiade.
1945c [1894] "La musique et les lettres." In *Oeuvres complètes*. Paris: Gallimard/Pléiade.

Marvin, Carolyn
1988 *When Old Technologies Were New*. New York: Oxford University Press.

Michel, François-Bernard
1984 *Le souffle coupé: Respirer et écrire*. Paris: Gallimard.
Nietzsche, Friedrich
1967 [1872] *The Birth of Tragedy*. Translated by Walter Kaufmann. New York: Vintage.
1980 [1883–85] *Thus Spoke Zarathustra*. In *The Portable Nietzsche*, translated by Walter Kaufmann. New York: Penguin Books.

Phillips, Lisa
1989 *Frederick Kiesler*. New York: The Whitney Museum/W.W. Norton & Co.

Proust, Marcel
1954 [1913] *Du côté de chez Swann*. Paris: Gallimard/Folio.

Saint-Amand, Pierre
1994 *The Libertine's Progress: Seduction in the Eighteenth-Century French Novel*. Translated by Jennifer Curtis Gage. Hanover, NH: Brown University Press/New England University Press.

Sami-Ali
1990 *Le Corps, l'Espace, et le Temps*. Paris: Dunod.

Schivelbusch, Wolfgang
1988 [1983] *Disenchanted Night*. Translated by Angela Davies. Berkeley: University of California Press.

Vasse, Denis
1974 *L'ombilic et la voix*. Paris: Le Seuil.

Valéry, Paul
1973 *Cahiers*, vol. 1. Paris: Gallimard/Pléiade.
1974 *Cahiers*, vol. 2. Paris: Gallimard/Pléiade.

Vidler, Anthony
1987 *The Writing of the Walls*. New York: Princeton Architectural Press.

Villiers de l'Isle-Adam
1983 [1874] *Contes cruels*. Paris: Gallimard.

Weiss, Allen S.
1989 *The Aesthetics of Excess*. Albany, NY: State University of New York Press.
1995a "Erotic Nostalgia and the Inscription of Desire." In *Essays in Sound #2*, 26–33.

1995b *Phantasmic Radio*. Durham, NC: Duke University Press.

Whitehead, Gregory

1992 "Out of the Dark: Notes on the Nobodies of Radio Art." In *Wireless Imagination: Sound, Radio, and the Avant-Garde,* edited by Douglas Kahn and Gregory Whitehead, 253–63. Cambridge, MA: MIT Press.

Wyss, Beat

1990 "*Ragnarök* of Illusion: Richard Wagner's 'Mystical Abyss' at Bayreuth." *October* 54:57–58.

Shards of Voice

Fragments Excavated toward a Radiophonic Archaeology

Alexandra L.M. Keller

Junk Mail: Or perhaps not, depending on what you think of babies. But this announcement has arrived in mail boxes across the world since the beginning of the century. They were all posted the same day, but (the mail being what it is) they continue to be received at irregular and wholly unpredictable intervals. There is no telling where or when. But the why is clear enough. This is the announcement of Radio Natal Day, and it reads,

Announcing
The Birth of Dead Air

Oh dear, is it too late to send a gift?

• • • • •

Radio and recording have changed everything for the voice. Before photography there was painting, and the visual image—the representational possibilities of the body—typically has been perceived as unfolding in a progressive, inevitable trajectory toward verisimilitude until the still camera, and then the motion picture camera, made abstraction possible. Painting, according to this logic, was freed from the bonds of representation and left to explore other issues. As flawed as this paradigm may be in itself, it is instructive in terms of articulating a quite different paradigm for the development of radio. For the representation of the voice (a separate concern from its mimicking by, for instance, a conch shell, a moaning wind, or any other musical or quasi-musical device) has from the start had an indexical status, like the photographic image. Yet also from the start, that indexical representation's potential for abstraction and altered reconstitution has existed. And in this divergent model we come to see the unique representational status of the voice as it struggles to describe itself as by turns separate from and integral to the body.

Consider the many voice-related idioms in the English language: "Cat got your tongue?" when we do not know what to say; "At the top of my lungs," when we are yelling as loud as possible; "Frog in her throat," when the vocal apparatus is not working properly. As often as not, vocal idioms speak around the very word *voice*, unwilling to name it. Even when they do not—as in, "I've lost my voice"—there is a certain instability. The voice, unlike other things that can be lost, such as sight, hearing, touch, and smell, is not a sense. Nor is it something that can be lost and replaced with a prosthetic, like an arm or a leg. (True, there are synthetic voice boxes, but in the voice's very evanescence—versus the presumed permanence of a finger—we may see how such devices do more to compromise identity than do limb replacements.)

The voice occupies a liminal status in relation to the body as a whole—as often neither/nor as either/or. Of what is a voice made? When the voice is not voicing, what is it doing? Does the voice have a role in the construction of silence beyond absence or restraint? Radiophonically speaking, the disembodied voice emphasizes all the more the irresolvability of its nature in relation to the body that produces it, and of which it is an essential, if contingent, component.

The remainder of this text strews itself out as shards collected from an archaeological site—a site consisting of past, present, and future. These remnants, real and imagined, are to be pieced together in any number of ways to achieve a legible form. Both more resilient and more malleable than ceramic or bone, these fragments submit themselves readily to recombination, speaking, in their continuous reassemblage, something of the heterogeneity of the prehistory of radiophony.

• • • • •

One has nostalgia for that which has been stolen, or given, or spirited away—that which is, in Antonin Artaud's intention of the word, "soufflé." Make no mistake, Artaud was not worried about some protracted *glissance de la voix.* Such slippage is glorious, but theft is terror. *La parole soufflé,* the word breathed is the word (always already) stolen. Speech, as the manifestation of word-breath, is always a treacherous process. To speak we must breathe. We must inhale and exhale. Or as the signs at any French gym will remind you, *inspirer* and *expirer. On inspire d'expirer. On souffle de mourir. La parole et la mort.* We are, in speaking, inspired to die. A conflation-translation which is itself breathing. We inhale French and exhale English. Speech: a rate of expiration with a date of expiration.

Such a little leap from *souffle* to *souffre,* just the fatiguing of a stiffness in the "l" to a stooped "r." Only a lexical bad posture to distinguish narration from its inherent pain.

• • • • •

White Noise: Nearly all voices of the dead speak in a single French idiom. Everything they say is *l'esprit de l'escalier*—the wit of the staircase—the words of retort and returned derision we wish we'd had at the moment of the inflicted wound. Such words fail us in that moment's present tense, and it is only in the past, as it becomes present to us while we beat a hasty retreat, that *le mot juste* becomes clear to us. It is worse for the dead, and when they come back to haunt us, they are unloading all those afterthoughts that have accumulated in the afterworld. The dead come back to say what they wish they had said. This is why their messages are often inscrutable to us, the living, for we do not have the benefit of the original context. So the meaning is lost to us, and in this loss a tragedy is repeated for the ghost. For if the living person forfeited the opportunity for a snappy comeback, the snappy comeback cannot even now be appreciated without the original insult.

And we wonder why they rattle their chains so insistently.

• • • • •

In the 29 December 1991 issue of the *New York Times Magazine*, there is an ad in the "Holiday Shopping at Home" section for a Voice Changer Telephone. The text begins, "Even your own Mom won't know it's you [...]." It continues to describe a state-of-the-art telephone which

> has another special feature: the ability to disguise you [*sic*] voice so that even your own Mom or your dearest friend won't be able to recognize you. You can pre-set the vocal pitch to any of 16 different characteristics—male and female. That can be a lot of fun, but it is also a first line of defense for women or for children who are left home by themselves.

The power of this item to do what it says is expressed in liminally cruel terms, since anyone inspired to purchase the voice-changer phone in order to alienate Mom or dearest friend might well be a person against whom a small child home alone would want to disguise her voice. But beyond that is the clear implication that any woman or child can protect herself by pretending to be a man. Inherent pragmatic chauvinism notwithstanding, the ad makes a radical suggestion—the voice can have transsexual capabilities.

Nevertheless, simply writing down in print this exceptional feature of the voice changer is not enough, even with an accompanying photograph of the actual item. Though the voice operates invisibly, it apparently needs to be expressed visibly—the voice needs to be portrayed as something one can see. So an ostensibly humorous cartoon accompanies the photo of the product (which looks like the average push button phone). It is an adjacent two-panel narrative whose inner frame is divided by a phone cord. In the first frame a man speaking on the voice-changer phone says to someone on the other end "Hello, John? This is Debra. You know, I think you're *really* cute—." Little musical notes surround the cartoon bubble, and in the background another man is holding his hand over his face and laughing. In the second frame, John's eyes are bulging out of his sockets, his right hand is clenched tightly around his own receiver, cartoon hearts are throbbing around his head, and in a bubble of imagination appears the image of heavy-lidded, glossy-lipped, and billowy blonde Debra.

In its visual representation, then, the voice is seen to widen its repertoire beyond the transsexual to include the transvestite as well as the homoerotic. The image and voice of Debra are virtual, the words and sentiment are actual, and the scenario (as representation) is real. And in sum, it gives the affirmative in the realm of the popular to the question audio artist Gregory Whitehead posed in a recent piece in which the human voice logs a stunning variety of synthetic modulations: "Do you want to have a voice like mine?"

• • • • •

Express Mail: The aesthetics of glossolalia are in many ways based on speed and duration. The listener's conclusion that the language is incomprehensible is at first drawn because the speaker is simply talking too fast and for too long to be understood. Or so it seems. With psychobabble, something on a par with synesthesia takes place in the listener's ear, a rearrangement of the characteristics of speech so that no aspect is entirely stable *qua* that aspect. For example, psychobabblers do not necessarily speak any more rapidly than "normal" speakers. But the fact that the untrained ear cannot always divide the steady stream of sounds into words, and that even when it can is unsure that the divisions are the proper ones, make the glossolalia seem to move faster, as if the speaker were a particularly adept auctioneer. And in moving faster, it also may appear to last longer. Likewise, volume, pitch, timbre, and so on may blend and switch, dip, do-si-do, allemande left or right, until the listener has managed to confuse herself as much as she has managed to be mystified by the babbler.

• • • • •

In the dream, Echo strolls through the boulevards of Hausmannian Paris. She does not speak, but telepathically sends interrogatives to passersby. They respond in her own voice, not as other people hear it, but as she does. Every word they speak resonates in her own larynx, and as one particularly loquacious gentleman inhales to begin a long discourse on the prettiest way to arrive at the Musée George Sand, she feels her own lungs puff up with air.

• • • • •

Barthes and Kruschev saw it the same way—future years ahead of them filled with nothing but the past. For a full lunar cycle, the deposed Soviet leader sat in a chair and wept. The future was terrifying because every day was identical to the last and next in its absolute emptiness, its utter lack of possibility. Kruschev did not even have a debilitating disease whose engagement with his corpus would at least mark the passage of time, decay being in this case a perverse sign of progress.

"I have only the grave before me," said Kruschev. To which, across the years and miles, Barthes sympathetically responded, "The future that remains to you [is] jail [...]. That is the definition of jail isn't it, when there is nothing new possible." Barthes was his own currency, and he had spent himself. Even these words, from a spoken appearance at New York University in 1978 entitled "Proust and Me," were copies of other words. Even Proust, whose own subject was the resurrection of the self to the self through the transformation of the sense-act into a speech/writing-act, was not enough for Barthes to delineate his present such that it boded a future.

Surely it was not a lecture, for as all who were there will tell you, Barthes was not reading but mourning. He was not speaking so much as bidding *adieu*, or in his case, perhaps, *au delà*. Surely it was not a lecture, for as all who were there will tell you, his death certificate is insufficient to the task.

Among Barthes's many, many dying words, he is said to have complained "that he felt decapitated, as if he were only a head." This is an extraordinary way to regard the paradigm of decapitation, especially for a French person. Decapitation is, after all, a word constructed on Latinate roots of taking away the head, and generations of guillotine souvenirs have ingrained in all European cultures the notion that when one is decapitated it is the body that loses the head and not the other way round. For Barthes, however, his body, once invaded by the interstates of the biomedical discourse, became an empty signifier, or even less than that. All that remained was his head.

The head/body split is the crux of the voice problem, and whether one thinks of the headless body or the bodyless head says much about from whence one believes the voice emanates. Decapitate from the chin up and the larynx remains with the body—so the head has no noise even if it mouths the words. Decapitate just above the clavicle and the body has no articulation—the noise has no form without the mouth.

Only one other soul in the universe would have understood Barthes, though she predeceased him by many thousands of years. Only Salome.

"Bring me the head of John the Baptist." Salome did not want his head as a caprice, and she wanted more of it than she got. When she did her dance of the seven veils for Herod, she engaged in a certain witchcraft, but nothing so quotidian as seduction. With the removal of each veil, Herod lost one of his senses.

First he could no longer taste the wine and dates on which he gorged because he could not sink his teeth into Salome's soft and hard thighs. Then he could not smell the incense that he had designed to replicate as closely as possible the smell of sweet sweat and blood emanating from her pubic region. (Little known about Salome is the fact that she menstruated constantly, every day of the year—or not at all when she preferred. It was part of her charm.) Then he could not feel the embroidered silk cushions underneath and all around him, so stitched to give him his best guess of tongue-coaxed goosebumps across her breasts.

(One need digress here into secret histories. Herod Antipas loved Salome more for her name than for her fleshly lures and snares, and he was certainly no match for her wit. Herod's own great-grandfather, Herod Antipater, had too been smitten with a Salome. Salome Alexandra, able queen of Judaea, mother of the last of the Maccabees, and widow of two brothers, was the object of a desire so suppressed that it buried itself deep in the genes. Herod was therefore as unable to separate himself from the desire for his niece, as one with hazel eyes—prosthetic devices notwithstanding—is unable to make those eyes truly brown.)

Another veil down and he could not hear the musicians he had assembled just in case—someday, oh, someday—Salome really would dance. Then he could not see the dancing Salome—kinetic dervish of a representation, his closest approximation to the Salome who feasted on his own flesh in his own dreams.

Then, when the five senses he knew he possessed had taken leave of him, a sixth veil dropped to steal the voice with which he called her name. Only then did he realize that the voice was indeed a sense, and if there were a sixth, well, then he surrendered to the loss of a seventh.

And it was in the lowering of the last veil, as it slit the close currents of midnight air, that Herod relinquished the sense with no name; and as he fell away from himself, his hard breath let go in the affirmative to the voice of his own DNA, which whispered in the *effleurage* of cherry-flavored, walnut-scented sign language, "Bring me the head of John the Baptist."

• • • • •

Radio Postcard/Radio Postscript (*"Wish you were hear"*): Driving in a car over long distances with the radio playing can be like slow-dancing with the aleatory. Try this: tune the radio to one station and leave New York City in a car driving due west on Interstate 80. Notice how the signal fades, overlaps with other signals, goes utterly dead, is interrupted by overflying helicopters, is taken over by ghost voices, flutter, and alien sounds, and so on. When you get to San Francisco, drive down the California coast and head east again, this time taking only local and back roads. See if the nature of the disruption is any different. See if you are any different.

In an alternative version of this experiment, try having a conversation with your radio as you drive through some of America's more breathtaking landscapes. How does your radio describe what's whooshing past it? How does your radio describe what *it* is whooshing past? Which one of you talks more?

Wired

Schreber As Machine, Technophobe, and Virtualist

Mark S. Roberts

Becoming Machine

> *In consequence of the many flights of rays, etc., there had appeared in my skull a deep cleft or rent along the middle, which probably was not visible from outside but was from inside. The "little devils" stood on both sides of the cleft and compressed my head temporarily to assume an elongated almost pear shaped form. The screws were loosened temporarily but only very gradually, so that the compressed state usually continued for some time.*
>
> —*Daniel Paul Schreber ([1903] 1988:138)*[1]

Daniel Paul Schreber, perhaps more fatefully than any 19th-century figure, was immersed—sometimes against his will—in a world of appliances, quasi-machines, devices, and mechanistic technology. He was, in fact, born and raised among appliances and devices. According to biographical accounts, Schreber's childhood was spent squarely in the midst of his father's various mechanical inventions, and, at times, he may have even served in the role of a guinea pig to actually test out these orthopedic and child-rearing devices.[2]

Paul's famous father, Moritz, seemed to have a talent—or, as some would argue, an obsession—for developing a whole range of what would commonly be considered orthotic appliances, devices intended to fit around certain bodily parts, and to improve everything from posture to mental attentiveness and toughness. These devices, profusely illustrated in Moritz Schreber's most popular book, *Kallipädie*, ranged from simple chin straps to head and back holders, and each contained an elaborate interlacing of leather straps and, sometimes, metal clamps intended to restrain, constrain, discourage, or improve unacceptable movements or postures. One might also imagine that Moritz Schreber, in his capacity as orthopedist, had an office full of various other prosthetic devices, such as artificial limbs, club-foot braces, crutches, back and neck supports, etc. In addition, one could well view Moritz the medical scientist, though still Kantian in his orientation, as emerging into a world of electric magnetism, synaptic charges, and cathexes, since descriptions of his specific medical practice and theories indicate that he was well aware of contemporary views of brain functions consonant with modern neuroscience;

that is, those focusing on the electromechanical aspects of brain physiology (see Lothane 1992:173–74). It is clear, then, that Paul's early orientation and the very environment in which he was raised was filled with quasi-mechanical and technological devices, intensified, one would assume, by the constant flow of patients creaking and clanking in and out of Moritz's combination home-office with an array of prosthetics, restraints, crutches, and braces.

Schreber's immersion in the growing techno-culture of the latter half of the 19th century would not be limited to his father's medical practice. As a gymnasium and university student in the late 1850s and 1860s, he would no doubt have been exposed to a broad range of the popular physical and biological scientific ideas of the period. For example, Alexander von Humboldt's multivolume work, *Kosmos,* was widely distributed during the 1850s, and it is estimated that no less than 80,000 copies of the work had been sold by the end of the decade. Among a number of other things, von Humboldt stressed the importance of mechanical inventions in the march of scientific revolutions—so much so, that he placed the invention of the telescope above virtually all other theoretical scientific discoveries (see Cohen 1985:260). His technical work in geography and meteorology—he was a pioneer in delineating isothermal lines—also deeply affected his conception of scientific progress. His technical genius, carried over into the popular *Kosmos* series, included the very early use of extremely sophisticated scientific instruments in a continuous survey in orography, geophysics, meteorology, earth magnetism, and even a nascent form of ecology. All considered, the emphasis von Humboldt and his followers placed on technological progress must have greatly magnified young Paul's already vivid impressions of the great force and significance of machines, mechanics, and technology, that is, almost on a global scale of determination.

Of course, von Humboldt's contribution was just one among many other advances made in medical, biological, and physical technology and theory during the period. With the contemporaneous work of the Helmholzian school of biophysics, for instance, we encounter a striking vision of a physical and mathematical mechanics applied to human perception, physiology, and, ultimately, to the entire life process. In a certain respect, Hermann von Helmholtz brought the precision of mathematical equation directly inside the human perceptual system, inside the head of the living organism. In his passion for mechanical reduction, Helmholtz, among other things, measured the optical constants of the eye with his own invention, the ophthalmometer, investigated the radii of the curvature of the crystalline lens for near- and far-sightedness, and even proceeded to apply the principles of electrodynamics to brain and nerve physiology. All this was, in turn, seen as possible because of the central theory of the conservation of energy which provided a "scientific" explanation for what was then viewed as troublesome metaphors about the mysterious essences of living things. With Helmholtz's purely mathematical expression of life functions, the

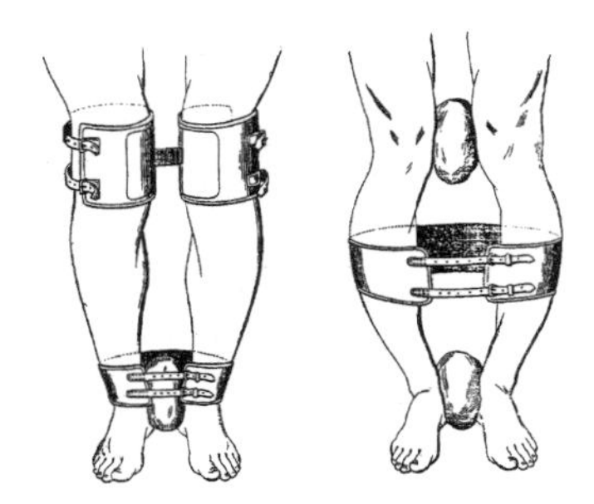

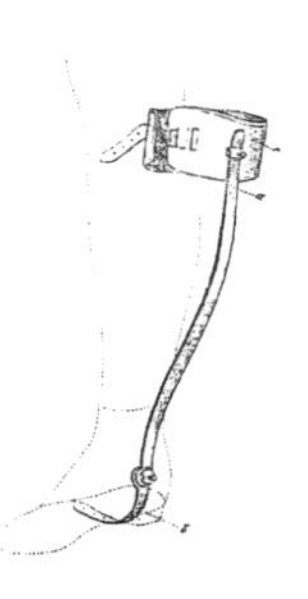
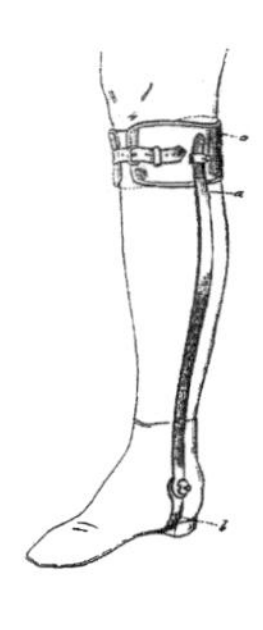

1. Illustrations of orthopedic devices from Moritz Schreber's Kallipädie *(1858:92 [left]; 205 [center]; 89 [right]). (Courtesy of Mark S. Roberts)*

organic functions could at last be seen as physico-chemical phenomena, leading to a mechanical explanation for the entire life process.[3] One of Herbert Spencer's American disciples, Edward Youmans, provides a vivid and decidedly rhapsodic description of this very process:

> Not only does it [the law of conservation of energy] govern the movements of the heavenly bodies, but it presides over the genesis of the constellations; not only does it control those radiant floods of power which fill the eternal spaces, bathing, warming, illuminating and vivifying our planet, but it rules the actions of and relations of men, and regulates the march of terrestrial affairs. Nor is its dominion limited to physical phenomena; it presides equally in the world of the mind, controlling all faculties and processes of thought and feeling. The star-suns of the remoter galaxies dart their radiations across the universe [...] and impressing an atomic change upon the nerve, give origin to the sense of sight. Star and nerve-tissue are parts of the system—stellar and nervous forces are correlated. (in Russett 1989:106)

The neurological extension of mechanics, alluded to by Youmans in the above passage, was perhaps best represented in German psychophysical science by Johann Friedrich Hebart and Gustav Theodor Fechner. Hebart, who died just prior to Schreber's birth (1841), developed a system of dynamic psychology which included a theory of unconscious mental processes and a conception of internal "forces" possessing specific "quantities." In fact, Hebart very simply defined psychology as "the mechanics of mind." In his passion for mechanical explanation in psychology, he even went so far as to postulate a mathematical formula for working out how much an idea was suppressed.

Fechner was even more rigorous regarding psychomechanics. Unlike Hebart, who made certain concessions to the Leibnizian and Kantian metaphysical traditions, he imposed what amounted to an absolute mathematical formula on the sensations, arguing that the magnitude of a stimulus could be measured by a given law: one must simply multiply the stimulus magnitude by a constant ratio. Hence, a stimulus of, say, 10, 20, and 40 ounces should yield equal degrees of sensation. Fechner further argued that the subjective intensity of sensation varies directly with the increase of the strength of a stimulus. He even worked out the difference between psychical increases (arithmetic) and physical ones (geometric) in terms of logarithmic equations, accounting for the differences between the two. This logarithmic relation would, in turn, account for the then troublesome "identity hypothesis," precisely because sensation had indeed been measured in a subject. Effectively, the human organism was nothing more than a mechanical operator, which could be measured by exact mathematical expressions, in terms of logarithmic relations between stimuli gradations.

The psychomechanics of both Hebart and Fechner, of course, inspired much of later-19th-century thinking in psychology and brain physio-anatomy. The figure who would eventually become Schreber's polymorphous nemesis, Paul Emil Flechsig, was one of those clinicians of the period who was influenced by this sort of thinking. Flechsig, even as early as Schreber's first hospitalization in 1884, was deeply concerned with the brain as a kind of pseudo-machine whose various parts and locations could be pinpointed in terms of specific functions. His theories regarding mental dysfunction always centered around organic etiology, and he stressed the absolute importance of the "brain mechanism's" relation to the entire organism:

> But it should be important to the physician that such psychological analyses are but a small part of his task, and in my opinion in no way the most

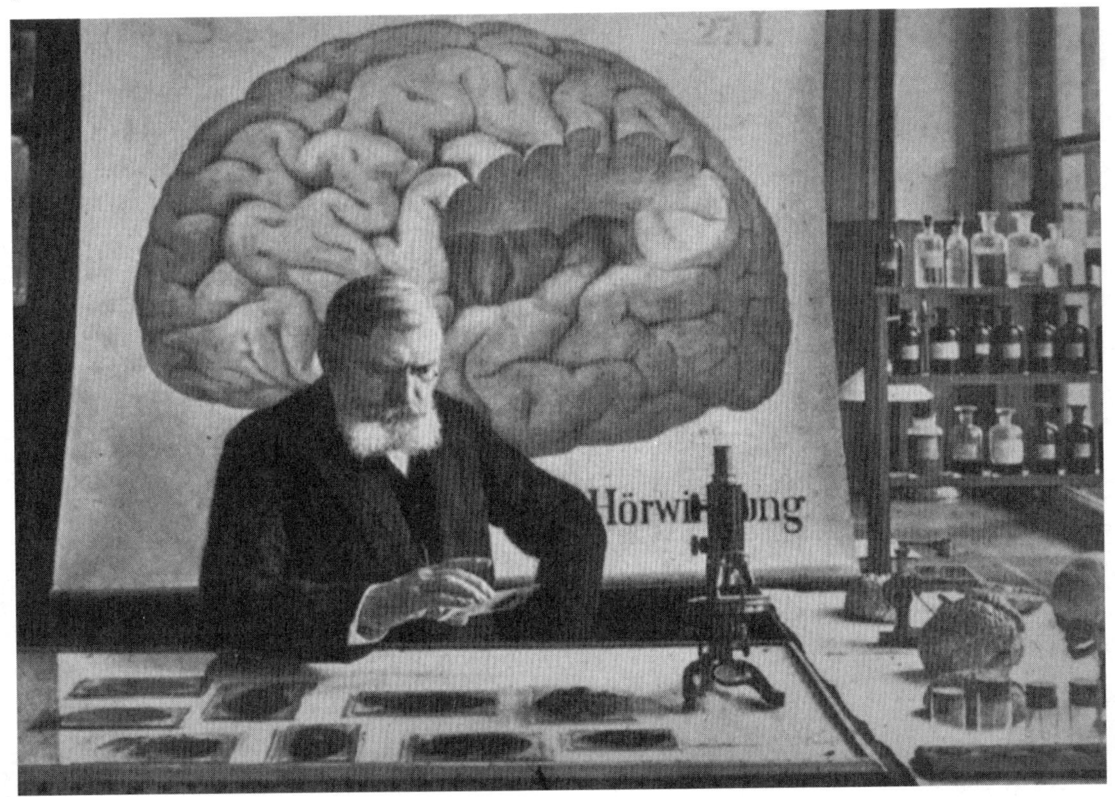

2. "All this was compounded, we can presume, by the dreadful sight of the jars of pickled brains that lined the walls in Flechsig's office, as well as the massive brain chart ensconced above his desk." (Photo in Karger 1909; courtesy of Gilles Fardeau)

> important. [...] The specific medical thinking begins only when the physical factors are kept in mind which are the cause of psychological changes. [...] The proper object of investigation is the localization and the nature of the underlying somatic processes or factors, completely in the spirit and meaning of modern scientific pathology—not more and not less. It is enough to demonstrate strong and lawful, even if remote, relations between the physical and the psychical. The exact knowledge of the brain mechanism and the entire organism is indispensable. (in Lothane 1992:211)

For Flechsig, then, the brain was, so to speak, a complex map, dotted with a multitude of loci, each of which, when pinpointed and fully understood, would yield some specific knowledge about abnormal behavior. It was perhaps with this in mind that he interpreted Schreber's second illness, in 1894, as a form of delusional paranoia—a disorder which Flechsig attributed to "diseases of the association centers and sensory centers" (in Lothane 1992:218). With this extension of psychobiology, psychophysics, and "brain mythology" to Schreber's early treatment and diagnosis, we find a progressively more intimate expression of mechanics in Paul's life, of a direct attempt to determine the precise point at which the psychophysical machine had malfunctioned. Schreber was not only tormented by fearful, desultory hallucinations, but his very treatment and the perception of his "disease" were now profoundly affected by a kind of psychomechanics, by a theory of the body and brain as an interactive machine that is moved by mechanical impulses and drives, and electromagnetic forces. All this was compounded, we can presume, by the dreadful sight of the jars of pickled brains that lined the walls in Flechsig's office, as well as the massive brain chart ensconced above his desk. Slowly, progressively, the devices he had experienced and was subjected to in his youth, the contexts of techno-culture, began to mesh with the very fibers of his being, with his mind and his body. Not—as Morton Schatzman (1974) and William Niederland (1974) have argued—as the sources of his frightful hallucinations, but, rather, as the central metaphor for what he would eventually become: a machine.

Schreber's progressive mechanization, his being rendered a machine, accelerated during his stay at the Sonnenstein asylum. His attending physician at the asylum, Dr. Guido Weber, like Flechsig, viewed mental dysfunction from a largely psychophysical perspective. He was neither original nor inventive in his view of clinical psychiatry, borrowing much from his mentor, Emil Kraepelin, and spending most of his time publishing and lecturing in the rather technical area of forensics. He was thus most comfortable in following a variety of rather straightforward psychiatric paradigms, and this to the extent of utter dogmatism (see Lothane 1992:271). Weber's mechanistic bias is revealed by his static and extraordinarily rigid view of the whole phenomenon of mental illness. On this, Weber himself writes:

> As colorful and inexhaustible the individual variations of cases of mental illness may be, as constant are the main outlines, and apart from the arabesques of the individual case the basic characteristics of the forms of mental illness are repeated with almost surprising, monotonous regularity. (in Schreber [1903] 1988:317)

The expression "monotonous regularity" itself gets repeated monotonously in almost everything written about Schreber during his years of hospitalization. The staff at Sonnenstein often described his general behavior as rigid and repetitive. Weber, in his report to the Superior Court at Dresden, described his demeanor in a certain instance as "pathological," as screwing up his eyes, grimacing, and holding his head in an extraordinary position, that is, as excessively mechanical and rigid (in Schreber [1903] 1988:324). He was repeatedly observed sitting in the courtyard at Sonnenstein, immobile and staring up at the sky. Almost like a radar antenna, he would mechanically turn his head from one side to the other, as if receiving some form of cosmic transmission. His famous "bellowing" and "howling" were also described in decidedly hydraulic terms: during conversations which proved to be stressful—for example, with his wife, Sabine—he would race off and release pressure with several good bellows or screams, returning to the conversation perfectly calmed. This release of growing pressure by bellowing was also observed on certain occasions when he was dining with guests at the asylum. Even his obsessive piano playing had something of the mechanical to it. It seemed that whenever he was excessively frustrated or disturbed he would "let off steam" by banging vigorously on the piano. This observed machinelike presence and behavior at Sonnenstein could perhaps best be summed up by an entry on his chart dated September 1895: "Often laughs loudly and piercingly and screamingly repeats the same words. From time to time stands totally still in one spot and stares at the sun and grimaces in a most bizarre way" (in Lothane 1992:295).

Schreber even suffered a certain degree of mechanization—albeit in retrospect—in the hands of Freud, who, oddly enough, was by far the most psychologically oriented of his early interpreters. Despite his rather classic psychoanalytical reading of Schreber's illness—the old unresolved Oedipus complex: passive homosexual fantasies leading to castration anxiety which, later, leads to a homosexual identification with his first doctor, Flechsig, etc.—much of Freud's analysis of the case draws upon markedly psychobiological and psychomechanical concepts. This tendency stems in part, according to Frank J.

Sulloway, from Freud's "biogenetic-Lamarckian presuppositions," which allowed him to attribute considerable traumatic force to what he called "pure fantasy" by tying these early fantasies to a "phylogenetic memory-trace" (Sulloway 1979:387). From this perspective, libidinal, or biogenetic, energy could be seen as having a specific mobility and quantity of force, and thus, like electric current, could be withdrawn, or redirected toward any number of object-choices. This phenomenon becomes quite clear in Freud's own explanation of Schreber's illness, that is, his supposed paranoid dementia:

> And we can understand how a clinical picture such as Schreber's can come about, and merit the name of paranoid dementia, from the fact that in its own production of wishful fantasy and of hallucination it shows paraphrenic traits, while in its exciting cause, in its use of the mechanisms of projection, and in its outcome, it exhibits a paranoid character. For it is possible for several fixations to be left behind in the course of development, and each of these in succession may allow an irruption of the libido that has been pushed off—beginning, perhaps, with the later acquired fixations, and going on, as the illness develops, to the original one that lies nearer the starting-point. (Freud 1953:113)

For Freud, then, Schreber's condition was the result of a libido fixation during the "narcissistic" stage of development—a stage occurring more or less midway between the libido's maturation from auto-eroticism to heterosexual object-choice. Schreber's libido just chose the wrong object—i.e., someone with the same genitals—and he spent the rest of his life regretting it, that is, forming elaborate defenses against the object-choice. The currents of libido, creating fantasies and forces beyond the subject's control, simply designate a path of maturation—in Schreber's case, one that leads ineluctably to a struggle against passive homosexual fantasies, a struggle marked by powerful and debilitating delusions. Given Freud's view, one might conclude that Schreber was, so to speak, "plugged into" madness—a view that differs in kind, but not so much in intent, from those proposed by others in the earlier psychomechanical and psychophysical traditions.

Machine and Technophobe

"Plugged into" madness, rendered into a machine, strapped into restraints, probed by devices, subjected to the psycho- and electromechanical theories of the time, Schreber was naturally both intensely aware of the fact that he had become a machine and horrified that he was one. His profound awareness is evident in the many colorful passages in the *Memoirs* that refer to his mechanization, his feeling—or as some would argue, his delusion—that he had become machinelike and was being "run" by someone or something. His fear of becoming completely mechanical—robotic—and his resistance to this transformation surface in a set of brilliantly inventive strategies intended to combat the repetitiveness and regularity of his treatment and his own experience and behavior.

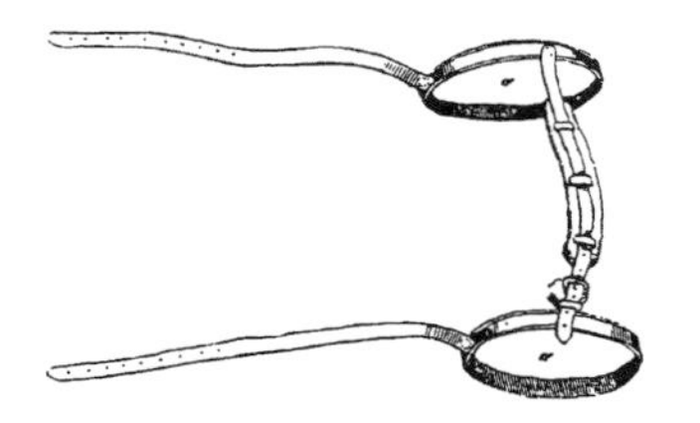

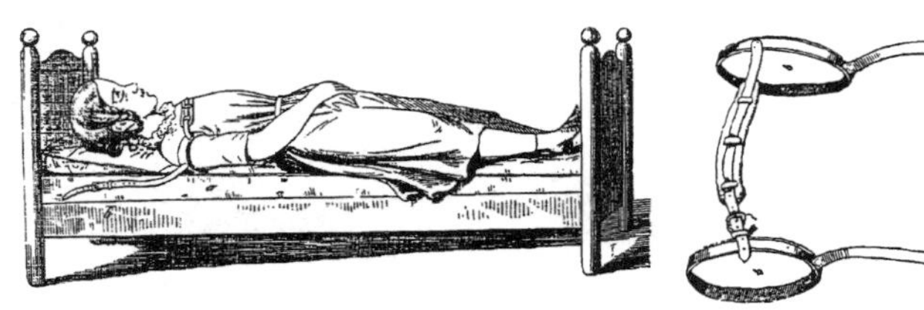

3. Illustrations of orthopedic devices from Moritz Schreber's Kallipädie *(1858:174). (Courtesy of Mark S. Roberts)*

The most obvious expressions in the *Memoirs* of electromagnetic forces and mechanics are those involving the "rays" (*Strahlen*) and their conduits, the nerves, and what Schreber calls "filaments." Schreber was fully convinced that certain types of filaments or wires were implanted in his body so as to make him receptive, as well as captive, to a variety of vocal messages carried by the rays. He describes the implanting of these "wires" in terms eerily close to the way some sort of radiophonic or telecommunicational device might internally view itself receiving signals from the outside:

> I see the same phenomena with my *bodily eye* when I keep my eyes open; I see these filaments, as it were, from one or more far distant spots beyond the horizon stretching sometimes towards my head, sometimes withdrawing from it. Every withdrawal is accompanied by a keenly felt, at times intense, pain in my head. The threads which are pulled into my head—they are also carriers of the voices—perform a circular movement in it, best compared to my head being hollowed out from the inside with a drill. ([1903] 1988:227)

The analogy of signal reception is extended by Schreber in the Postscript section of the *Memoirs*, where he attributes his inordinate ability to hear barely audible "cries of help" to a phenomenon like "telephoning":

> I even believe I have found a satisfactory explanation of why cries of help are only audible to me and not to other people [...]. It is presumably a phenomenon like telephoning; the filaments of rays spun out towards my head act like telephone wires; the weak sound of the cries of help coming from an apparently vast distance is received *only by me* in the same way as telephonic communication can only be heard by a person who is on the telephone, but not by a third person who is somewhere between the giving and receiving end. (229)

Solar and cosmic transmissions (the "rays") are further described down to the specific configurations of their patterns: they do not arrive in a straight line but rather in a circle or parabola—which, by the way, are forms that certain wave and particle transmissions sometimes assume. When they do arrive, Schreber argues, they must be "slowed down by some mechanical means; otherwise they would simply shoot down into my body, drawn to it by the enormously increased power of attraction [...]" (288–89). This description, once again, could apply to any number of modern receiving devices. Schreber's emphasis on "slowing down" the rays with some "mechanical device" is uncannily close to how, for instance, a television receiver operates. Normally, the signal is received as a radio frequency broadcast and then fed into a transformer which converts it into synchronizing pulses that in turn drive the cathode ray tube and the loudspeaker system. The conversion from radio frequency broadcast to synchronizing pulses—what Schreber refers to as "slowing down"—is essential to the entire process, since without it, the initial radio waves would remain undifferentiated.

When discussing the exact nature of the "rays," Schreber even tends to characterize them in purely electromechanical terms. Although the rays carry a considerable amount of information, Schreber suggests that they are "essentially without thoughts" (*die Hauptgedankenlosigkeit*). By this he means that the rays are often without memory, devoid of any thought, and therefore of any specific human or divine intentionality. Without thoughts of their own, the rays are simply intermediary devices, intended to convey ideas and information in a wholly detached way—in a way remarkably similar to how artificial intelligence works. The various microcircuits of a computer, for example, store, carry, access, calculate, process, etc., information, but really don't have any thoughts of their own; they electronically translate the information entered by an operator. What is ultimately conveyed is electronically produced bits of information, not thoughts or intentions. A phenomenon not unlike that associated with the "rays-being-essentially-without-thoughts":

> But the nerves without thoughts must also speak in order to slow down their approach. As they however lack thoughts of their own, and as there are no beings with thoughts of their own at the places (on stars, celestial bodies) where they are loaded with poison of corpses (one may picture these beings which are also responsible for the writing-down system either as human shapes like the "fleeting-improvised" men, or in some other way) the quiescent totality of divine rays can (when they approach) only give them or drum into them to speak what they have read as my own undeveloped thoughts. [...] This is the rough picture I formed of the thousandfold repetition of the rays "being-essentially-without-thoughts." (235–36)

Closely associated with "the rays," "the voices" are also characterized in electromechanical terms, in terms of a curious "prerecorded" language that operates apart from actual speakers, including Schreber himself. In the *Memoirs*, Schreber frequently mentions that he is being pursued, constantly taunted, by "voices." These voices, however, are somewhat unusual—even in this context!—because they continually repeat expectable things and, in the end, become "mostly empty babel of ever recurring monotonous phrases in tiresome repetition" (139). The reasons for this are complex, at least as far as Schreber is concerned. The voices, Schreber insists, issue from several sources or media, but one of the main sources is the "talking birds." The birds are remnants (single nerves) of souls of human beings, which carry with them a particular "tone-message" associated with their respective human souls. The tone messages are learned by rote, and therefore merely repeated without either feeling or sense. Indeed, given Schreber's description, the words seem to be composed out of purely vibrational sound elements: "I cannot say how their [the birds] nerves are made to vibrate in such a manner that sounds spoken or more correctly lisped by them sound like human words" (167). The voices of the talking birds, then, are not really voices in the sense of human speech at all, but are, rather, the tonal equivalents of spoken words; or, one might suggest, analogous to the way prerecorded sounds are experienced by the listener—not human, but the mechanical tonal equivalent of human speech. Is it live, or is it Memorex?

The sense of audiotape or even photographic recording is also invoked by Schreber when he describes "tests" applied to his apprehension of certain terms, phrases, or visual objects. According to him, God has taken to "examining" him whenever he is in a state of "not-thinking-of-anything." In short, the procedure consists in God causing people around Schreber to say certain words by stimulating their nerves. An example of this would be a madman throwing in a certain learned term that he had remembered from the past—perhaps a foreign word or two. The terms themselves "come to Schreber's ears" accompanied by the phrase "has been recorded," which is directly "spoken" into his nerves. This procedure serves to assure that Schreber has indeed heard the phrase, and, ultimately, to test whether he understands it. This can also occur with visual phenomena. Like a camera, whenever Schreber observes certain objects, the phrase "has been recorded" resounds in him:

> For example, when I saw the doctor my nerves immediately resounded with "has been recorded," or the senior attendant—"has been recorded," or "a joint of pork—has been recorded," "railway—has been recorded" [...]. All this goes on in endless repetition day after day, hour after hour. (188)

Obviously, there is no shortage of electromechanical analogies in the *Memoirs*, as Schreber goes on at great length describing his frightening transformation from man into machine—an idea that had marked parallels outside the walls of Sonnenstein, in the late Romantic rejection of the Enlightenment and the long-standing concern with the dehumanizing consequences of the Industrial Revolution. But what is perhaps even more interesting than this transformation is his attempt to resist it, to reiterate his basic humanity and his sense of worth and self-dignity. There are a number of strategies he used to accomplish this, most of which employed some type of countermeasure to the monotony and regularity of what I have here characterized as his progressive mechanization.

One of the numerous strategies utilized by Schreber to free himself from mechanization consisted in his obsessive, seemingly unending conflict in the *Memoirs* with the supposed "soul-murderer," Flechsig. From the very beginning of the autobiography itself, which corresponds more or less to the first days of Schreber's second illness and his treatment at Flechsig's University Clinic, Schreber had identified Flechsig as a kind of "machine-master," as an evil "soul" who held extraordinary power over virtually every physical and mental move he made, or, perhaps more accurately, was compelled to make. Hence, in order for Schreber to be truly human, to be free from the monstrous influences imposed upon him by Flechsig's will, he had to ultimately defeat his evil machinic persona.

This mastery Flechsig exerted over his mind and body began, it appears, with a central act of physical repression—that is, at the point at which Flechsig's rather nasty attendants dragged Schreber from bed and brutally abused him, tossing him about, pinning him on a billiard table to restrain him, and then finally throwing him into a cell (66). Following his incarceration, Schreber was further punished by continual isolation at the clinic; a number of restraining drugs were also administered. The drugs, according to Schreber's account, were often "forced down his throat." As his suppression and isolation became progressively more severe, his feelings of being "used," abused, and of being generally dehumanized also increased. Finally, Schreber tried to resolve the whole problem of dehumanization by taking matters into his own hands, by employing the only human means he saw possible to escape the horrors of the depersonalizing, degrading, repetitive, painful experience of Flechsig's clinic: suicide. In a manner of speaking, he wished to "unplug" the machine so as to counter Flechsig's pernicious influence:

> Completely cut off from the outside world, without any contact with my family, left in the hands of rough attendants with whom, the inner voices said, it was my duty to fight now and then to prove my manly courage, I could think of nothing else but that any manner of death, however frightful, was preferable to so degrading an end. I therefore decided to end my life by starving to death and refused all food. (76)

Schreber soon learned that more subtle means would work against Flechsig's insidious control. The divine rays, which carried the voices that sometimes attacked Schreber, were originally seen as—at least in part—being under Flechsig's control: "The only possible explanation I can think of is that Professor Flechsig in some way knew how to put divine rays to his own use" (69). When he was attacked by the voices carried by the rays, however, unlike the attacks of Flechsig's "rough attendants," he tried to counter their colorless force by reciting the poems and prose pieces of the two greatest writers of the German Romantic tradition, Goethe and Schiller, or reeling off words and phrases he had consigned to memory. In the face of compulsive thinking (*Denkzwang*) and the monotonous cacophony of the voices, carried by the rays, he repeated his own inventions, his own unique aesthetic memories to shut out the mechanical drones (see Weiss 1988:75).

These defenses against the droning "voices" and a variety of other mechanical forces, however, go far beyond the fundamental conflict with Flechsig, extending to a whole series of depersonalizing crises. For example, Schreber often resisted the evil "nerve" forces that threatened his reason and humanity by invoking his own erotic sensations, or what he termed "soul-voluptuousness" (*Seelenwollust*). Simply stated, soul-voluptuousness was the result of a colossal work of God in which Schreber was to be transformed into a woman, "unmanned," so as to serve the higher purpose of procreating a future race. This process included the replacement of his masculine nerves with feminine ones. Although this at first occasioned considerable consternation on Schreber's part, he ultimately began to realize that his progressive feminization had an up side: intense voluptuousness. He eventually learned to turn this voluptuousness to his own use, to, among other things, employ it as a counterreaction to any attempt to depersonalize him, or, as he would put it, to "rob him of his reason." When, for example, he was confounded and attacked by senseless phrases sent to him through his nerves, he was able to render these phrases more palatable and personal by invoking the forces of soul-voluptuousness. He also observed that the nerves themselves were less foreign, less "terrified" when they entered his body, due to the pleasant sensation of soul-voluptuousness: "But the attraction lost all its terror for these nerves, if and to the extent they met a feeling of soul-voluptuousness in my body" (Schreber [1903] 1988:150). Effectively, his own transformed erotic feelings were able to give him a sense of autonomy, of being—at least on some occasions—in control of what he estimated to be "hundreds of thousands" of celestial nerve-visitations.

Along with the colossal struggle with Flechsig and soul-voluptuousness, the so-called "bellowing miracle" (*das Brüllwunder*) served as still another weapon against Schreber's transformation into a machine. From the onset of his second illness, Schreber was observed as having the peculiar habit of loudly bellowing and screaming. He seemed to do this at odd intervals, at a variety of times, and on differing occasions, which led his doctors and attendants to believe that the so-called "miracle" was strictly compulsive and without much design. Schreber, however, thought differently. He saw a double purpose in the bellowing: first, to create a "representation," an impression of someone who is demented, and, second, to "drown by bellowing the inner voices" (166).

Schreber does not really explain what he means by the first purpose of the bellowing—that is, creating an impression of a demented person—but he does on several occasions discuss the second purpose. As we have seen, Schreber felt that the "voices" attacking him were one of the main sources of his terror and depersonalization: "[...] my nerves cannot avoid the sound of the spoken words; the stimulation of my nerves follows automatically and compels me to think [...]" (174). His entire sense of his own sanity, his very inner peace and being were continually destablized by the constant, repetitive roar of the inner voices. Like an engine that one cannot shut off, the voices repeat the most elementary, demeaning, repetitive, whirring phrases and sounds imaginable. Hence, the only way that Schreber could find any peace, and, in turn, any sense of his own autonomous being, was to "drown out" these sounds; to scream from the depths of what remained of his own personality, to vocally overwhelm the whirring and mechanical hum of the voices. As he himself says:

> [...] I have to allow the bellowing as long as it is not excessive, particularly at night when other defensive measures like talking aloud, playing the piano, etc., are hardly practicable. In such circumstances bellowing has the advantage of drowning with its noise everything the voices speak into my head, so that soon all rays are again united. This allows me to go to sleep again or at least to stay in bed in a state of physical well-being [...]. (228)

So Schreber, in the face of becoming an anonymous machine, of being intricately "wired," of having his every perception and apprehension "recorded," of being continually victimized by uncontrollable "voices," etc., felt compelled to develop a set of strategic defenses to counter this mechanization. In doing so, he was able to establish his own personal identity up against what amounted to nearly a lifetime of depersonalization and mechanical usurpation. The recalled flashes of his father's technical devices, strapped around patients, pictured in books; the nascent technoculture in which he developed; the psychophysics and electromechanics of his treatment—all of these, for the moment, would disappear in the intense feelings of self he was able to invoke through these clever endopsychic strategies and defenses. Effectively, Schreber had, for the moment at least, saved himself from the unbearable lightness of a droning, machinelike being.

Virtualist

In Schreber's zeal to penetrate the depths of his mind, to struggle with the hellish inner voices and "miracles" that constantly beset him, he, unknowingly, provided a remarkable, futuristic glimpse into yet another electromechanical phenomenon: virtual reality. Although virtual reality—being virtual—is rather difficult to accurately define, most modern virtualists agree that it entails the creation of some type of hyperrealistic simulation. Howard Rheingold, in his expansive book on the subject, defines the term broadly as "a computer generated artificial world" (1991:1). By this he means that human consciousness is augmented, or perhaps more accurately, mimicked by the instrumentality of a computer. The process typically involves the creation of a completely convincing illusion that one is immersed in a world that actually exists only inside a computer. The so-called "virtual traveler" is customarily hooked-up visually to the computer through an interactive helmet, which is in turn extended by a VR (Virtual Reality) input glove. The VR input glove—like the new body input hookups—virtually "grabs" virtual objects in virtual space; that is, it allows one's visual perceptions and motor responses to navigate through virtual space.

The effects created in virtual space, in VR, are, as might be expected, both spectacular and mind-boggling. A typical virtual voyage might consist of the following:

> My consciousness suddenly switched locations, for the first time in my life, from the vicinity of my head and body to a point about twenty feet away from where I normally see the world. The world I saw had depth, shadows, lighting, a look of three-dimensionality to it, but it was depicted in black and white. [...] Twenty feet away from my body, my view of the world changed in response to my physical motion. I began to accept the odd sensation that accompanied the act of transporting my point of view to that of a machine—until I swiveled my head and looked at myself and realized how odd it seems to be in two places in the same time. What you don't realize until you do it is that telepresence is a form of out-of-the-body experience. (Rheingold 1991:255–56)

Sound familiar? It should. There are any number of passages from the *Memoirs*, particularly in those sections devoted to "miracles," that correspond almost exactly to this sort of virtual excursion. Schreber's notion of "picturing" (*Zeichnen*), for example, rests on the assumption that man retains all memories, by virtue of impressions on his nerves, in the form of pictures in his head. These pictures can then be "looked at in the head" by the rays. But, quite re-

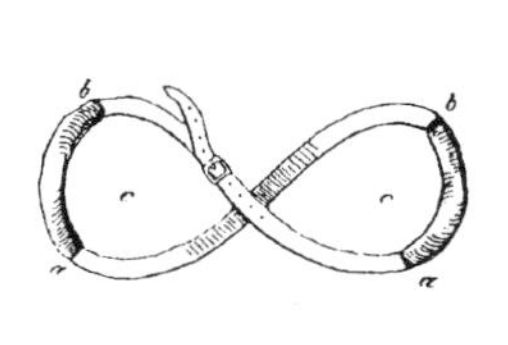

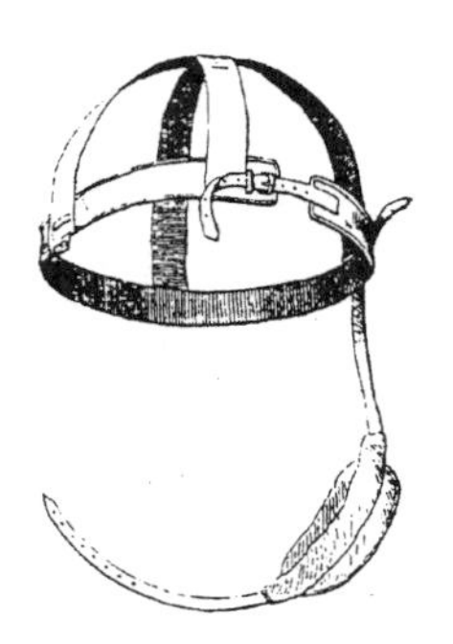
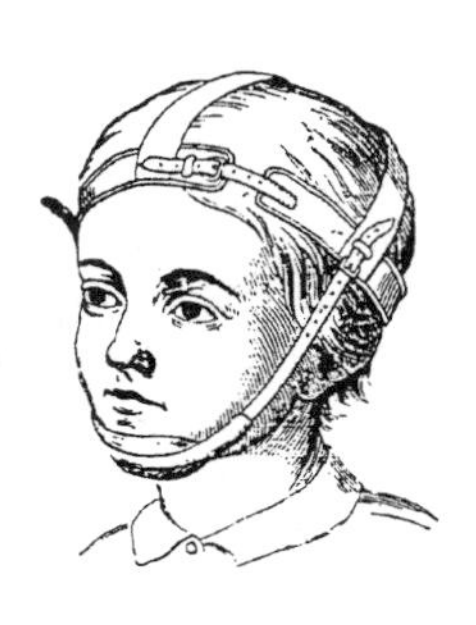

4. Illustrations of orthopedic devices from Moritz Schreber's Kallipädie *(1858:198 [left]; 220). (Courtesy of Mark S. Roberts)*

markably, they can also be watched by the rays outside the head, in what one might conceive of as some sort of fantastic "theatre" of the rays. Indeed, Schreber's description of the occasional shift of the "pictures" from inside to outside his head bears an uncanny resemblance to the futuristic world portrayed in the above-mentioned VR voyage:

> [T]hese images become visible either inside my head or if I wish, outside, where I want them to be seen by my own nerves and by the rays. I can do the same with weather phenomena and other events; I can for example let it rain or let lightning strike [...]. I can also let a house go up in smoke under the window of my flat, etc. [...] I can also "picture" myself in a different place, for instance while playing the piano I see myself at the same time standing in front of the mirror in the adjoining room in female attire [...]. In the same way as rays throw on to my nerves pictures they would like to see especially in dreams, I too can in turn produce pictures for the rays which I want them to see. (Schreber [1903] 1988:181)

Among the many unusual phenomena characteristic of the hyperreal simulations of VR, we can also count what is generally referred to as "tele-existence." This is a state in which machine-sensed data can be transformed into human-sensed data, so that the operator can experience a symbiosis with some robotic or telemetric device. Effectively, the technology is intended to give humans greater knowledge of and control over reality (existence) through complex electronic devices, as, for example, sensing helmets are used in military aircraft to enhance the technical efficiency of fighter pilots. Tele-existence, then, is, among other things, a means of amplifying human experience and perception so that they can interface with the very complex, barely perceptible world revealed by electromechanical instrumentalities.

Schreber, remarkably enough, also refers to something quite similar to tele-existence in the *Memoirs*; it turns out to be germane to what he calls "miracles concerned with damaging my body" (247). According to Schreber, he had, during his years at Sonnenstein, suffered an extraordinary number of physical catastrophes: tearing headaches, accelerated breathing, several varieties and colors of plague, penile retraction, diminution of body size due to contraction of the thigh bones or vertebrae, a different heart placed in his body, pulmonary phthisis, lung worm, smashed ribs, "Jew's stomach," torn or completely vanished intestines and gullet, suppression of the "seminal cord," putrefaction of the abdomen, muscle paralysis, etc. The only problem concerning the diagnosis and, ultimately, medical science's belief in the actual existence of these dire symptoms, is that no one could find a device sophisticated enough to pick them up, to record these miraculous occurrences. In a manner of speaking, Schreber's afflictions required a radically new kind of measuring device, one that would function something like tele-existence does in VR. He says as much when he laments upon the difficulties one would encounter in making a thorough examination of his body:

> If it were possible to make a photographic record of the events in my head, of the lambent movements of the rays coming from the horizon, sometimes very slowly, sometimes—when from a tremendous distance—incredibly swiftly, then the observer would definitely lose all doubt about my intercourse with God. But unfortunately human technique has not yet the necessary apparatus for investigating such sensations objectively. (248)

Finally, there is a parallel in the *Memoirs* to what VR people humorously refer to as "teledildonics." The term "dildonics" was coined to describe a machine invented by a hardware hacker, How Wachspress, which converts sound into tactile sensations. The erotogenic possibilities of the machine naturally depend upon where on the body the consumer wishes to attach it. But VR researchers see bigger things than mere self-stimulation for this sort of technology, and future visions of teledildonics can be enormously intriguing:

> Now, imagine plugging your whole sound-sight-touch telepresence system into the telephone network. You see a lifelike visual representation of your own body and of your partner's. Depending on what numbers you dial and which passwords you know [...] you can find one partner, a dozen, a thousand, in various cyberspaces that are no farther than a telephone number. Your partner(s) can move independently in the cyberspace, and your representations can touch each other, even though your physical bodies might be continents apart. [...] If you don't like the way the encounter is going [...] you can turn it all off by flicking a switch and taking off your virtual birthday suit. (Rheingold 1991:346)

Schreber, as one could easily guess at this point, was well on his way to developing a 19th-century version of teledildonics—at least, within his own head. Without going into excessive detail, one of the most obvious forms of this sort of futuristic sexual encounter occurs with his notion of "soul-voluptuousness" and its relation to the "rays" and "nerves." If one recalls, the very state of soul-voluptuousness usually involves some "telemetric" contact with the nerves ("nerve-contact") of certain divine individuals, including, but not restricted to, figures like God, God-Flechsig, Ariman (the lower God), Ormuzd (the upper God), and even, sometimes, the fleeting-improvised-men (*flüchtig hingemachte Männer*). The reason these figures occasionally "phone in" or get "on-line" is the intense erotic sensations existing in Schreber's body, the "soul-voluptuousness." For instance, whenever God became aware of Schreber's strong feminine sexuality, he would be attracted to his body by virtue of an intensified feeling of eroticism that was transmitted through the fila-

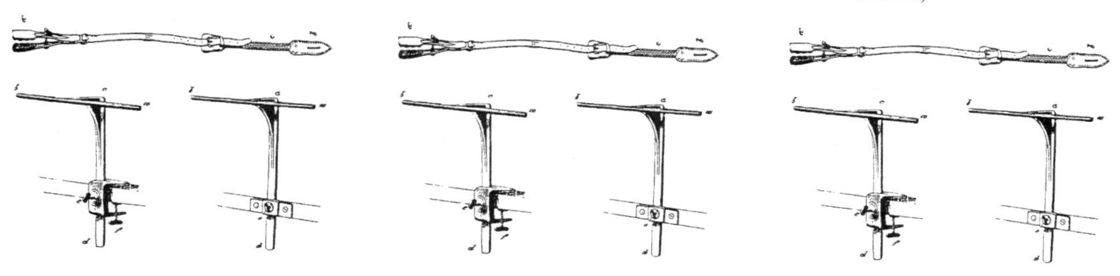

5. Illustrations of a head holder (top) and a back "brace" from Moritz Schreber's Kallipädie *(1858:199 [top]; 203). (Courtesy of Mark S. Roberts)*

ments connecting Schreber with God. It was as if God's sensual representation was relayed to Schreber as a result of the intense erotic feelings sent through the nerve-contact, and vice versa. Both mutually shared a kind of "hyper-spatial" communication in which sexual desire, sensuality, perhaps even orgasmic pleasure were exchanged. In discussing one of these encounters with the lower God (Ariman), Schreber recalls precisely this sort of erotic, hyperspatial partnership: "The lower God (Ariman) as stated, did not object to losing himself with part of his nerves in my body, because he almost always met soul-voluptuousness there" (Schreber [1903] 1988:150).

"There," of course, being the most private recesses of a man who, when summing up his titanic struggles, emphatically claimed that his ultimate end would be to rest in a state of permanent blessedness and soul-voluptuousness—states that are perhaps becoming the two greatest virtualities of modern life (252).

Notes

1. Daniel Paul Schreber was born in Leipzig in 1842. His father, Moritz Schreber, was a well-known educator and orthopedist, and his work is still popular in Germany today. Paul studied law and eventually went on to become an important jurist, rising to the position of *Senatspräsident* of the Saxony Appeals Court in Dresden. In 1884 Schreber suffered an illness that was diagnosed as hypochondriasis. After being treated by Paul Emil Flechsig—a neurologist of some note at the time—he returned to his duty as judge. In 1894 he had a much more severe breakdown, and was subsequently sent to the mental asylum at Sonnenstein, where he spent nine years. During his incarceration, Schreber wrote his now famous *Memoirs of My Nervous Illness,* which was eventually published in 1903. It was the *Memoirs* that served as the source for Freud's proxy analysis of "The Schreber Case," which was published in 1911. Since then, there have been numerous articles and books written on the case. These works have, over the years, led to lively, sometimes acrimonious, debates about the case and the man. Schreber died in a state of severe mental and physical collapse in the Leipzig-Dösen Asylum in 1911.
2. In his generous comments made on an earlier draft of this paper, Zvi Lothane had suggested that Moritz Schreber was not concerned to develop machines in the conventional sense of the word, nor was he interested in experimenting with mechanical devices, which, in retrospect, is consistent with his own thesis in *In Defense of Schreber.* I fully accept this suggestion. However, my intention here is not to establish factually that Moritz Schreber actually invented "machines," but, rather, to draw a loose sketch of the young Paul Schreber growing up in the midst of what would at least appear to a young boy as an unusually mechanistic environment, and would thus serve as a prelude to his "becoming machine."
3. Regarding the biological and evolutionary sciences, Schreber makes reference in a footnote in the *Memoirs* to a number of texts that he had already read prior to his illness (see Schreber [1903] 1988:80, fn. 36).

References

Cohen, Bernard
1985 *Revolution in Science.* Cambridge, MA: Harvard University Press.

Freud, Sigmund
1953 "Psychoanalytic Notes on an Autobiographical Account of a Case of Paranoia (*Dementia Paranoides*)." In *Standard Edition,* vol. 12. London: Hogarth Press.

Karger, S.
1909 "Festschrift für Paul Flechsig." *Monatsschrift für Psychiatrie und Neurologie,* 26.

Lothane, Zvi
1992 *In Defense of Schreber: Soul Murder and Psychiatry.* Hillsdale, NJ: The Analytic Press.

Niederland, William G.
1974 *The Schreber Case: Psychoanalytic Profile of a Paranoid Personality.* New York: Quadrangle.

Rheingold, Howard
1991 *Virtual Reality.* New York: Summit Books.

Russett, Cynthia Eagle
1989 *Sexual Science: The Victorian Construction of Womanhood.* Cambridge, MA: Harvard University Press.

Schatzman, Morton
1974 *Soul Murder: Persecution in the Family.* New York: New American Library/Signet.

Schreber, Daniel Gottlieb Moritz
1858 *Kallipädie.* Leipzig: Fleischer.

Schreber, Daniel Paul
1988 [1903] *Memoirs of My Nervous Illness.* Translated by Ida Macalpine and Richard A. Hunter. Cambridge, MA: Harvard University Press.

Sulloway, Frank J.
1979 *Freud: Biologist of the Mind.* New York: Basic Books.

Weiss, Allen S.
1988 "The Other As Muse." In *Psychosis and Sexual Identity: Toward a Post-Analytic View of the Schreber Case,* edited by David B. Allison, Prado de Oliveira, Mark S. Roberts, Allen S. Weiss, 70–87. Albany: State University of New York Press.

HeadHole

Malfunctions and Dysfunctions of an FM Exciter

Christof Migone

Quietly, let's unplug everything, blindfolded. Ears plugged, nose clamped, tongue tied, let's strip the hardware off radio. The exciter, heart of the FM transmitter, comes last. Once off, the purring winds down and we find ourselves radiophilic without transmitter. Commonly, the resultant dead air spells anguish and panic. No longer can the signal signify the voice and radiate it with power. For the moment, however, we will dwell and even revel in the air dead. There are remnants and potentials in our voices hereto untapped that will be sufficient to carry this broadcast home.

The regular hosts from the Center for Radio Telecommunications Contortions have decided to invite some special guests as cohosts for this program: the Analphabête—an unbridled voice; the Wireless Wired—a posthumous telephone; and the Transpiring Transistor—a translator exercise.

Radio without Transmitter

Radio voices are dead on arrival. Upon entrance to the studio, they are trained to sync lips to accepted tastes and are delivered to the listener excited, but hardly less than moribund. There are several factors working to stultify this delivery. Foremost is the monopolizing of radio space as carrier of information. "Fact: radio as wallpaper" (Moss 1990). Voices on radio are well combed and articulated, not masticated or salivated. They have been air-dried and dehydrated. "After seventy successful years in the wallpaper business, radio has many of the powers to flatten, smooth out, disembody and trivialize the information it conveys" (Moss 1990). It is no surprise that radio as a creative tool is still strange territory. There is a molding of the voice that standard radio requires; a predetermined format shapes the voice to its well-treaded contours. The mold is defined by a blandness that is crass, or, alternately, a crassness that is bland. The cast of the voice is now an immutable crutch.

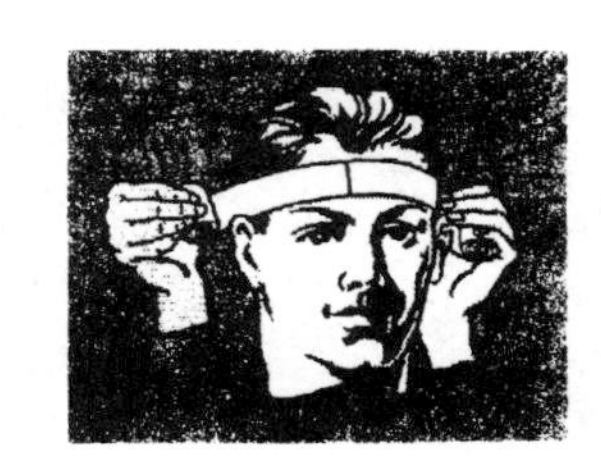

Even college and community radio stations cannot pretend to be free of this sort of ossification. They have become a viable alternative and in so doing have suffered no small amount of institutionalization.

In a text entitled *Speech: A Handbook of Voice Training, Diction and Public Speaking*, Dorothy Mulgrave states that "the well adjusted speaker will conceal from his [*sic*] audience any sign of tension or discomfort" (1954:133). The result is radio where everything is false and nothing is permitted.[1] The concealment of vulnerability is how the game is lost. Learning to hide can occur either via a subtle transformation or an overt mutation. All who possess a remunerating voice have come across alliteration exercises and articulation jaw breakers. One such exercise tape in diction repeats the mantra: "Learning to speak well is an important and fruitful task" (*Learning* n.d.). The tape then enumerates the voice types: (1) Neutral Voice, (2) Raised Larynx, (3) Falsetto, (4) Creaky Voice, (5) Whisper, (6) Whispery Creak, (7) Whispery Voice, (8) Whispery Falsetto, (9) Creaky Voice, (10) Creaky Falsetto, (11) Whispery Creaky Voice, (12) Whispery Creaky Falsetto, (13) Breathy Voice, (14) Harsh Whispery Voice, (15) Tense Voice, (16) Lax Voice.

We can only speculate what an inflation of the numbers will produce: (795) Breathy Tense Creaky Neutral Whisper Voice... (9,676) Wounded Raised Larynx Lax Vitriolic Falsetto Voice... and, finally, (126,789) Creaky, Breathy, Radiated, Harsh, Tense, Electrocuted, Fondled, Neutral, Contorted, Raised Larynx, Throated, Vexed, Whispery, Transpired, Articulated and Vehiculated, Incontinent, Vagabonded, Phantomized and Phased, Jaundiced, Relayed, Postdetermined and Postdigital, Deregulated, Mellifluent, Erased, Manipulated, Fast-forwarded, Battery-operated, Synoptic and Phatic and Tonsilitic, Glottal and Colossal, Salivaphile and Expectorant, Lecherous, Licentious, Projected, Reverberated, Remote-controlled, Vivisected, Transistorized, Modulated, Masticated, Animated, Assiduous, Analphabête Voice.

It is inflation, amplification, exaggeration from the HeadHole *in* to the HeadHole *out* that brings the voice alive. Stelarc once stated with a maniacal laugh that "the problem isn't getting it in, the problem is getting it out" (1993). He was trying to get a sculpture out of his stomach. We are merely trying to let it out of the mouth. Once the voice is excavated via the mouth, you get recognized by strangers. *I know you, I really know you... I know who you are* (Migone 1996).[2] The type of recognition alluded to here is the trademark of the Analphabête and is based on nothing more than a sharing of the air dead. A particular kind of kinship, where to unlearn language is the common gene, where the bel canto is infected by the pleasure of imperfections, and where nowhere is where it's at.

David Moss once pronounced radio dead (1990). This leaves the field open for resuscitations: radio is dead, long live radio. "I've always hoped and fantasized that radio was the perfect medium in which to propagate subversive artistic activity, by its very normalcy radio would function as a sort of culture dish of art bacteria and it would grow its own audience" (Moss 1990). Conversely, the radio petri dish is also growing its own artists. Its size cannot accommodate the transmission apparatus nor does it need to, audience and artists having spawned from the same bacteria. The politics that espouse that everybody should arm themselves with a transmitter can now make the leap to the following scenario: radio without transmitter. Perhaps this is the required script to trigger the postdigital age. Skip the digits' demands for detected errors and corrected codes. Skip the automation which "looks empty but sounds full" (Oakwood n.d.). Fast forward to the postdigital age, an age with a taste more savory than the antiseptic and a time beyond the accelerated. An age where the Analphabête will be spoken and heard through every orifice.

The Dead Line

> *Ken Charles Barger, age 47, accidentally shot and killed himself last December in Newton, North Carolina. Awakening to the sound of a ringing telephone beside his bed, he reached for the phone but grasped instead a Smith & Wesson .38 special, which discharged when he drew it to his ear.* (Hickory Daily Record)[3]

Telephones are so much a part of our lives. They have also become part of our deaths. Few of us, perhaps, face the spectacular pathos of Ken Charles Barger's final call. Picking up the phone is dangerous. The phone is loaded and has become a terminal shock treatment connecting us to the beyond. In deadication to Mr. Watson—the first caller to the transcendent—we are suffocated by our own information web.

The telephone is increasingly without answer, increasingly without voice. You can navigate through the myriad choices of your touch-tone phone until you turn blue, a shade prior to expiring. *Larry King Live* is in fact *Larry King Dead*. Before the show was canceled, all callers to the syndicated North America–wide program should have, once on air, shouted "King Asphyxia," then reveled in some choking sounds, and finally hung up. Thus, we introduce the posthumous telephone and its users, the Wireless Wired, not as phoenix rising but as living dead: "at once grotesque and familiar, banal and exaggerated, ordinary and on the edge" (Shaviro 1993:83).

Conversations are no longer the telephone's principal calling card. What we connect are transactions and suffocations. Mimicking Barger, a caller reaches for the phone but grasps instead an ad: "Let my voice blow you away" (Phone Sex 1993). Enticed by the thought of a *petite mort*, the caller then dials 1-900-xxx-xxxx, credit card at the ready. The little death ejaculates and the receiver gets messy. The telephone, as a panophonic device where sex and death are currency, engenders calls that blow you away.

Bruts

Selected Writings

He said you speak softly now and you don't want to speak out loud kelanelstikikosti postirmaisi secret police of madmen secret police also to prove that the makalam of promakalam prokalastarrokalaremsbrokelaisstrrmmakalaistostemarlokerster melaokester copy for me what you said a national sick person Monsieur in the end he said that the brain registers just that yesterday we were really persecuted but here we're still at least among madmen who says Jacqueline why do you always speak of something else when I answer you well was it to pass the time because I can no longer Cyrano de Bergerac Bergerac out loud.

—Jacqueline (in Oakes 1991:165, translated by Sophie Hawkes)

Ive lost tutch with mye selph Ide bee blyged iff yoo cood spair a pockit ahnker cheef for Jools Duddin. who evry wonn thynk Stynk. Tooby Plucky saye Hang onn Eye cood Boase toff havvin skraipt throo bye the

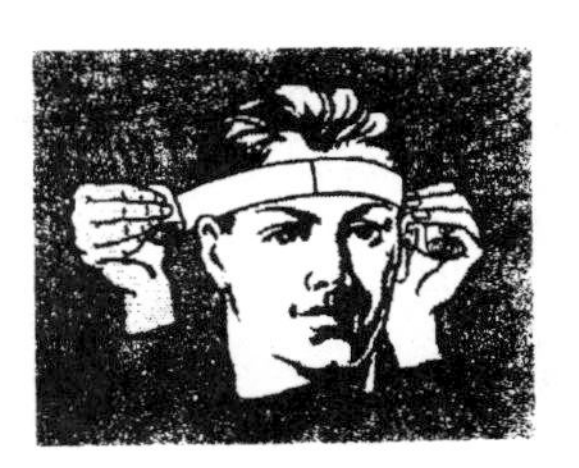

> Observing that many of his fellow citizens have cellular phones but are seldom called, entrepreneur Joachim Benz of Frankfurt-am-Main, Germany, has launched Rent-A-Call. For a five-mark fee each time, he will call subscribers on their mobile phones whenever they want him to; hundreds of clients have signed up. (*The Independent* 1994)[4]

It is only a matter of time before these subscribers suffer the fate of Ken Charles Barger. They should all be called and, once on the line, the caller should shout "King Barger is dead," then revel in some choking sounds and never hang up. After all, happy to be finally wired and wireless they will be touched. Reach out and _ _ _ _ someone. With this evidence, it seems that if being "live" is what we desire we cannot afford to wait for the call. Answering the phone is dangerous. We have to *make* the call. Obsessively, randomly, desperately, absurdly, let the phones ring! Let us begin with the progenitors. Glenn Gould's line is busy. We'll try later. Laurie Anderson has Moholy-Nagy on the other end. Avital Ronnel is conferencing with Martin Heidegger and Alexander Graham Bell. Try again later.

> Across the sleeping city two strangers are locked in anguished intimacy. Their lifeline is the slender thread—a coil of telephone wire. Minute by minute two strangers are joined in desperate communication. One is a man fighting bitterly to save a life; the other is the woman...who wants to die. (Silliphant 1965)

Sydney Poitier and Anne Bancroft play the two tangled strangers in the film *The Slender Thread.* Their telephone link negotiates life and death, and the coil's tension provides the movie's modus operandi. The apparatus seems to relish its role as bearer of bad news, as the anguished intimacy is heightened by the invisibility of the exchange. The telephone locks us in a state of imminence. "It was a wrong number that started it, the telephone ringing three

skninner mye teath. They diss pize mee they shitten mee. Iff yoo confined any whun nose smudgers mee bough tittall juss sendem upp too seamy inn the king derm Of heavy earn Thattle semmy upper bitt Nuffter grabber thatt littall Dar lingh jest gonn parst.

—Jules Doudin (in Oakes 1991:127-28,
translated by Roger Cardinal)

Emile Hodinos Josome - Son of Emile Jean Hodinos - Locked up in Ville Evrard for no good reason. Moulder - Modeller - Inventor - Typesetter - Compositor - miniaturizer - Engraver of Medallions. Baths - Showers - Straightjackets - Several Baths During the years 1877-1878-1880-1881-1882-1883-1884-85-86-87-88-89-90. During the years 1891, 1892, Bathing suspended. Bathing regularly each week. During the years 1893-1894-1895-1896-1897 approximately - Fed - watered - clothed at the Ville Evrard. Make my bed quite regularly. Empty out my chamber pot. Drawing - Writing - The - Brushing of my shoes - Cleaning my Combs - Hair fallen out.

—Emile Josome Hodinos (in Oakes 1991:133-35,
translated by Roger Cardinal)

times in the dead of the night, and the voice on the other end asking for someone he was not" (Auster 1985:6). Sorry, wrong number. The advent of features such as call display, call trace, and call block render one's phone a veritable fortress. A phone number is not only one's property but has become inextricably tied to one's identity. I am calling you. Sorry, wrong person. Dialing a wrong number: trespassing with a blindfold. Each dial could be the last, thanks to the suspense of imaginary barricades.

Cellular phones encourage traffic torticollis and remain economically prohibitive for most. Their cells have not yet undergone the inevitable diminutive plunge under skin. The inevitable, however, is near. With phones internally installed, they would become genuinely cellular and we would become answering machines. "Answering Machines: they are patiently training us to think in a language they have yet to invent" (Ballard 1992:277). With global telegeography grafted to our neurons, the phone companies would drive us insane, dialing up truly cellular damage. Ultimately, the Ma Bells would collude to have our offspring conceived and subscribed simultaneously. Bewildered, at first we would be answer machines with only questions. Once the telemicroscopic device becomes fully assimilated, we will be seen in the street talking to invisible people. Even the cellular-clad today can be likened to the so-called insane who wander our public streets having private conversations out loud. This type of activity is very much akin to the radio transmitter's functioning. You diffuse something that any (passersby) can pick up if so desired, while talking to an intended audience (the person at the other end of the wireless line). Proto-FM behavior, radios without transmitter. Gould's line is still busy.

The telegeography of the Wireless Wired is reaching radiophonic proportions. The call you place is broadcast in invisible ink all over the air on its way to its intended destination. Prince Charles calls with his pants down. "Let my voice blow you away." The cell phone flaunts one's private parts. "The art of being everywhere" is the same as the art of being nowhere (X 1993) *...well, why did they hang up? I have no one to talk to.* From 1988 to 1994, my radio program *Danger in Paradise* on CKUT-FM in Montreal regularly left callers hanging. Some of these calls were eventually weaved into the audio work, *Hole in the Head.*[5] Recounting telephone stories has particular poignancy on the radio: it multiplies the holes in the head until buoyancy is in question, and you start to sink. From porous heads, protruding holes welcome the apparatus and we mainline, the telephone as capsizing neuroma. As with any self-respecting tumor, it is malignant. We have become one of the Wireless Wired in the vein of Gene Hackman in the Francis Ford Coppola film *The Conversation*, Mark E. Smith in the Fall's song "Totally Wired," George Brecht in the performance *Three Telephone Events*, or Liam O'Gallagher in his audio work *Border Dissolve in Audiospace.* Just as the eavesdropper is precariously crouched on the eave as it drops, the Wireless Wired is strategically placed to become part of the conversation, even if there is nothing to say. "The answering machine is a democratizing instrument, but the kind of democracy that results is a rather odd one—a democracy in which everyone has an equal right not to participate" (Rosen in Kelly 1988:27).[6]

The nonnarrative narrative in the radio/telephone space is a symptom and a cause. The calls made in *Hole in the Head* were never answered.[7] The mere fact they were made weaved the narrative. *Ay ay ay, it's lonely out there in the middle of nowhere.*[8] "Telephones: A shrine to the desperate hope that one day the world will listen to us" (Ballard 1992:269). The narrative traces the attempts at making the connection. A narrative following the Wireless Wired on a line jagged and drunk, where en route *I wonder who you are. I wonder where you are. I wonder whether you are. If you are, can you are, be you are, are you be?*[9] The choice is in making the call the act and the site for an implosion of identities. An implosion where your story is no longer contained in a hermetic inner voice but is porous with strings and leads, weaving the mix.

The dead line ends the story by hanging up, leaving a trail of voices mixed up in a perplexing web. The neurotic posthumous telephone threatens our ears. Why call in *viva voce* when you can whisper directly to a synapse? Once atrophied, the wilted appendages would perhaps be saved from amputation in order to lay idle as backup generators of audio stimulation, should telephony malfunction or dysfunction. The ear is an appendix in the history of the technological human. All the while, the abundantly porous Wireless Wired is hardwired to a call which converses in perverted verse and remains ambiguous to the ears' fate. The noise of the ring is deadening, off the hook.

Bruts and Bruits[10]

> *Most writers write in order to make a sound, even if their tree is falling in a forest where there's no one to hear it. (Anshaw 1991:66)*

Writing is a sound in silence; a sound is struck in the same manner a letter is scrawled. Writing is often the expulsion of a compelling inner force, a drive to create, which does not necessarily occur in conjunction with the drive to communicate. The sound of the tree is silent because it is not heard. This implies sound has significance only at the point of reception. The *Bruts* writers are strangers to the literary establishment and indifferent to the rules of grammar (see Thévoz 1979). Ecrits Bruts is an extension of the term Art Brut coined by artist Jean Dubuffet. This art is produced by people outside of the art world, people socially and mentally marginalized (hence sometimes the term Outsiders is used). They write but they are not writers. Hence, we can place them outside the culture of reception. They belong to the culture of the insane, where creativity is more likely to be seen as symptomatic of a "condition."

As one might guess these writings navigate the borders of sense (see *Bruts* sidebar). But you don't have to be crazy to write non-sense; I am not proposing a pedestal of insanity. The propensity of the writings for alliteration, solecisms, spoonerisms, catachreses, jabberwockies, anagrams, ellipses, pleonasms, and portmanteaus all result in a prosodic chaos which invariably brings to mind sounds along with meanings. The reader acquires ears. What we hear are the sounds of our imagination interpreting the text, a process which exists in all reading to a certain extent. The Bruts writings, however, seem to really pick up the instruments and bang raucously away. Forget minuets, welcome the Nihilist Spasm Band.[11]

The Bruts writings are authored by people who spent the better part of their lives in institutions. By and large, their silence is rather a silencing, not a choice, but a sentence. Although dissimilar to a Cageian silence in terms of intent (in that sense Cage's silence is but a privilege), it is nevertheless "a silence full of noises" (Cage 1981:210). They write with abandon, because we have abandoned them. They are the embodiment of the Zen saying, "It is not the case that someone who is silent says nothing." The texts are embodied, because we have straightjacketed them. These texts are bodies. They scream, masturbate, contort, fuck, defecate, digest, exercise, cough, sweat, etc. They are not necessarily or exclusively loud or scatological, but they are undeniably tied to the individual rather than untied and strapped to the body of the institution of literature. Operating with this distinction, the *bruts* become *bruits* (for noise is also an outsider, as it is usually considered to be the opposite or even the negation of music). The texts are not meaningless for the noise they

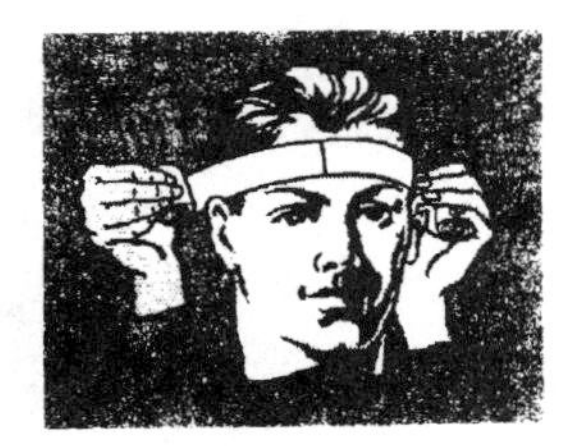

emit and their status as outsiders to literature, but are part of what Jean-Jacques Lecercle terms the *remainder:* "The relation of grammar and the remainder is one not of opposition or inversion but excess" (1990:60). Thus, the remainder is language at work on the delirious construction of accidents. It is the living language rather than its prescribed version. They are the jokes, the puns, the rants, the hallucinatory ramblings which play with, or even "do violence to," language (Lecercle 1990:60). The remainder is not an *other*, nor a marginalized obscurantism for, as Lecercle would assert, it is the method by which language becomes.

> There is always something grammatical about delirium, there is always something delirious about language. [...] Language is material not because there is a physics of speech, but because words are always threatening to revert to screams, because they carry the violent affect of the speaker's body, can be inscribed by it, and generally mingle with it, in one of those mixtures of bodies the Stoics were so fond of. (Lecercle 1990:58, 105)

Artaud's relevance is obvious at this juncture. "Artaud's terror was dark, filthy, emanating from the deepest recesses of his body, a body which his discourse tried, always unsuccessfully, to rejoin" (Weiss 1990:58). The body language in this case finds its recess in the anal tongue. A fusing of muscles which sparks the violence of the transgressor. A compulsive excess in vehement opposition to anal retention. The retentive is now a common analog to depict those who willingly institutionalize themselves in constructs (the Moral Majority being an obvious example).

For us as readers/listeners, the approach to these texts must be without caution. The passive is pacifier; it numbs. It is anal retentive. The entry point is the act of interpreting/translating. For the reader to develop ears is only an introductory step; as Gregory Whitehead wittingly points out, the ear is "just another hole in the head" (1991:141). Furthermore, in this case we are referring to sounds from the silent text. They are not heard, they are thought. It's the noise of the brain that I want to amplify.

> *yomart te i no te i no stat i o e cel chioz i zi vi vi zian vientse i e i niotsel e vi vi* (Artaud 1978:132)
> [you mart tea i no tee i no state i one cell choice i zip viva Zion Viennese Eli nix tell a vivid]
>
> *broita sen erver brait esa orzin erva brait osa orzin ervu* (218)
> [brute seen revere brain sea erosion sea brat Rosa ursine revue]
>
> *maloussi toumi tapapouts hermafrot. emajouts pamafrot toupi pissarot rapajouts erkampfti* (11)
> [malaise tumid tappets permafrost. emaciated jowls permafrost toupee passport rap jousts errs camp fit] (translation by author)

The above translations of Artaud are interesting because they are not translations per se, they are machinations.[12] They do not pretend to find universal meaning in a hermetic language but rather intrude, corrupt, and disarticulate the original. There is a certain paradoxical faithfulness in this approach, for it does not strive for accuracy, nor does it fabricate a neutral voice toward a *literaturization* of the embodied text. Translation connotes a professionalism which perpetrates this chimera of objectivity. *Traducson* is the French term for translation via sound and perhaps a more appropriate term for us to try to apply here. Lecercle's *The Violence of Language* provides us with an edifying example of traducson in the story of Leonora (1990:72). A native of the French West

Indies, where creole was forbidden at school, she instinctively retranslated the Latin from the Bible she had to learn by heart. "Ave Maria" became "lave lari la" (*lavez la rue là*: wash the street); and "miséricorde" became "mizire kord" (*mesurez la corde*: measure the rope). While it is through sound that the creolizations were performed, there are poetic correlations between the two ends. So, to isolate sound as the new currency of exchange is not satisfactory. It would signify the replication of the institutional model; it would become the norm. Where the tried and known fail us, and for the sake of irreverence, let's adopt the term "Transpiring Transistor" for a trial run as the new translator.

The Transpiring Transistor is naughty by nature and always noisy. It performs a kind of reading/listening that is inseparable from writing/voicing. In its desire for a herniated body of text it utilizes all senses to ferret its subjects. So writing is not a sound being silenced, just a sound in silence. And the sound in silence can be a noisy affair.

With regards to the Bruts writings and radio, Michel Thévoz, director of the Collection de l'Art Brut in Lausanne, Switzerland, makes the link effortlessly:

> Radiophonic expression is of particular interest and perhaps more akin to the Bruts writings than intimate communication. I am suspicious of intimate communication and its mirror effect. The effect is of a mutual complicity which tempts us strongly to subjugate ourselves into the image the other has of us. It is an aspect of conjugality which does not favor expression. I believe that love has never inspired any poet, it is rather conflict, confrontation, and jealousy which reveal Proust's genius and not affectionate sentiments. Thus, expression is truly freed from the constraints of this complicit intimacy when it can address more anonymous subjects.
>
> I have the impression that through radio one can be less susceptible to prejudice and more one's self. It might be paradoxical to have such a public medium the site of more honesty, more nakedness but as I said being in close proximity is constraining and stifling, thus perhaps the act of addressing none in particular might free up tongues (as was the case with Artaud). The Bruts writers had no access to radio but I would imagine that a microphone would have interested them greatly. They are often characters of parade and spectacle who invent a public for themselves—in the theatre the public is too present whereas with radio the guarantee of an audience without the usual face-to-face confrontation would surely have stimulated them. (1992)

In *Hole in the Head* I placed calls to the Bruts writers, and in the routing I, as a novice Transpiring Transistor,[13] found a simultaneous series of cacophonous stammers, sentences, and screeches emanating from my mouth. The Bruts microphone connected the delirious voice to the discordant radio. Thus, the calls (although unanswered) transpired and became the noisy affair that a silent text can be.

The Naked Parade

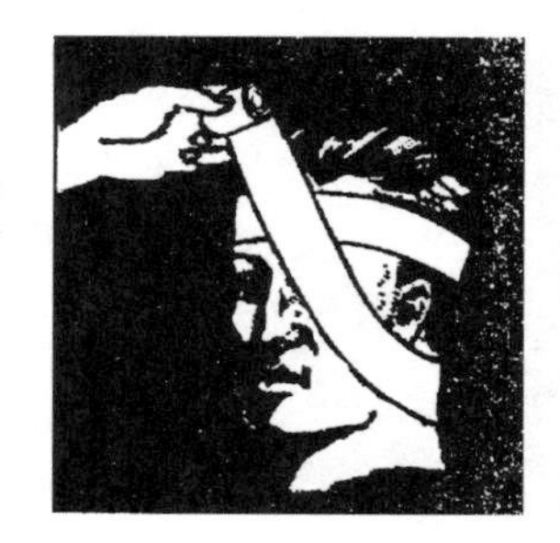

The FM exciter gives the audio signal the frequency of modulation, the rest of the transmitter boosts the signal and gives it power. When the exciter is not excited, we are bored.

> Today, the artist knows he [*sic*] can actually express himself [*sic*] *less* than others. Always and forever every day, he [*sic*] probes the elusiveness or the absence of expression which, if it manifests itself, does not reveal

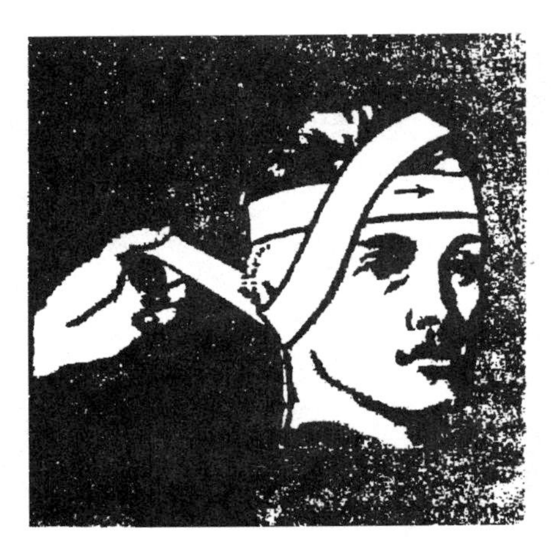

> itself *in* him [*sic*] but leaves *to* him [*sic*] the bitter task of giving it voice. (Paolini 1992:75)

Today, we find that we emit continuously, even in silence. The danger is neither in what we speak nor in what we hear but simply in what we fear. Nietzsche dubbed the ear as the organ of fear (Nietzsche [1881] 1982:143). While I would contend that fear is more pervasive and awaits at our every pore, the ear does function as an early detection system, and the more naked we are, the more it comes in handy.

As the HeadHole heals, the guest hosts fade with ears nevertheless perked and mouths resolutely vociferous. Rapidly, let's find the loose wires and reconnect the studio. No longer naked or dead, the machinery hums merrily oblivious to the damage done. The Analphabête makes an encore appearance, promptly losing the radio station's license to broadcast. The Wireless Wired is diligently learning how to solder intimate memories to neural synapses. And the Transpiring Transistor reaches for Louis Wolfson's *Le Schizo et les Langues* and never returns.

Notes

1. A paraphrase of William S. Burroughs's famous dictum "Nothing is true, everything is permitted" (1967:15).
2. Unidentified caller from audio work *HeadHole* (1996). Original call came from the program *Danger in Paradise* on CKUT-FM in Montreal, hosted by Christof Migone.
3. Uncredited news item, 21 December 1992. Quoted in *New Statesman & Society* (1993:47).
4. Uncredited news item from *The Independent*. In *Globe & Mail* (1994: A20).
5. *Hole in the Head* is a radio work that was commissioned by New Radio and the Performing Arts for the 1992 season of the New American Radio series. It is forthcoming as a CD release from Ohm editions, Quebec.
6. From "An Equal Right to Inequality: The Sociology of the Answering Machine," by Jay Rosen. In *ETC: A Review of General Semantics*, quoted in Kelly (1988:27).
7. The calls were made (metaphorically) to the *Bruts* writers (see "Bruts and Bruits" section and sidebar).

8. Unidentified caller from audio work *Hole in the Head.* Original call came to the radio program *Danger in Paradise.*
9. Gregory Whitehead from audio work *Hole in the Head.*
10. This section is a revised and expanded version of an article that appeared in *Sub Rosa* (1992:3).
11. Name of a music group based in London, Ontario. In the printed info for their 1984 cassette, *1984*, they state, unequivocally, "We are always loud."
12. My translations are inspired by Jocelyn Robert's "Art against Temperance," a text produced by running the "Front de Libération du Québec" (FLQ) manifesto through the English spellcheck of Microsoft Word. An excerpt from Robert's text: "Five le Quéebec limbered, five legs camaraderie priestliness poleitiques, five la revelations queebecoise, five Art against temperance" (1994:312).
13. The accompanying *TDR* CD features "Emile Josome Hodinos"; the piece is part of the Transpiring Transistor series of works derived from *écrits bruts.* Several pieces of this series—"Sylvain Lecoq #1," "Samuel D.," "Sylvain Lecoq #2," "Henri Bes," and "Henri Müller"—appeared on the CD *Radio Rethink: Art, Sound, and Transmission* (1994).

References

Anshaw, Carol
1991 "Divine Madness." *The Village Voice,* 23 July:66.

Artaud, Antonin
1978 *Oeuvres complètes* vol. XIV, 11. Paris: Gallimard.

Auster, Paul
1985 *City of Glass.* New York: Penguin.

Ballard, J.G.
1992 "Project for a Glossary of the Twentieth Century." In *Incorporations,* edited by Jonathan Crary and Sanford Kwinter, 269–79. New York: Zone.

Burroughs, William S.
1967 *The Exterminator.* San Francisco: Dave Haselwood Books.

Cage, John
1981 *For the Birds.* Boston: Marion Boyars.

The Independent
1994 "Social Studies." Michael Kesterton, ed. In *Globe & Mail,* 15 March:A20.

Kelly, Kevin, ed.
1988 *Signals.* San Francisco: Harmony Books.

Learning to Speak Well
n.d. Diction exercise cassette. Source unknown.

Lecercle, Jean-Jacques
1990 *The Violence of Language.* New York: Routledge.

Migone, Christof
1992 "Bruts and Bruits." *Sub Rosa* 2, 3:3.
1993 *Hole in the Head.* Privately published cassette.
1994 *Radio Rethink: Art, Sound, and Transmission.* CD. Banff, Canada: Walter Phillips Gallery.
1996 *Headhole.* On *Voice Tears, TDR* CD. Cambridge, MA: MIT Press.

Moss, David
1990 Speaking at the panel "New Identities for Radio" during the National Alliance of Media Arts Centers (NAMAC) Conference, Boston, 18 May.

Mulgrave, Dorothy
1954 *Speech: A Handbook of Voice Training, Diction and Public Speaking.* New York: Barnes & Noble.

New Statesman & Society
1993 "Forteana: Pythons, emus and peanuts can be the death of you." Paul Sieveking, ed. 3 September:47.

Nietzsche, Friedrich
1982 [1881] *Daybreak*. Cambridge, MA: Cambridge University Press.

Oakes, John G.H., ed.
1991 *In the Realms of the Unreal: "Insane Writings."* New York: Four Walls Eight Windows.

Oakwood Audio, Inc.
n.d. Advertisement for the "Media Touch" system, which automates radio stations.

Paolini, Giulio
1992 "Identikit." *Artforum* March:75.

Phone Sex
1993 "Let My Voice Blow You Away." Advertisement from the "Adult Personals" section of the *Montreal Mirror*, 26 August.

Robert, Jocelyn
1994 "Art Against Temperance." *Canadas*, special issue. *Semiotext(e)* 17:312.

Shaviro, Steven
1993 *The Cinematic Body*. Minneapolis: University of Minnesota Press.

Silliphant, Stirling
1965 *The Slender Thread*. Film, directed by Sydney Pollack. Paramount.

Stelarc
1993 Radio interview with Kelly Hargraves. CKUT-FM, Montreal, 5 April.

Thévoz, Michel, ed.
1979 *Ecrits Bruts*. Paris: Presses Universitaires de France.

Thévoz, Michel
1992 Interview with author. Lausanne, 1 December.

Weiss, Allen S.
1990 "K." *Art & Text* 37:56–59.

Whitehead, Gregory
1991 "Radio Art Le Mômo: Gas Leaks, Shock Needles and Death Rattles." *Public* 4/5:140–49.

X, Robert Adrian
1993 *Zero—the art of being everywhere*. Graz: Steirische Kulturinitiative.

Graphics
All images are from the Canadian edition of the St. John Ambulance First Aid Manual, 1964.

Cat's Cradle

Susan Stone

Aural outpost.
Frame linguistic.
Virgin image,
splayed, acoustic.

Ordered through
the discipline
of filings.

Consecration through reception.
Zealous practice of pressured perception.

What gives?

Magnetic deception, in reverse.
Forward, sounds much less perverse.

Tendril prehensile,
length discreet,
through a transport wending.
Acoustics pour,
a whiskey neat,
the glottal flow portending.

Head is laid upon the block.
Then, pressed, tone is the issue.
Pinned, noise gets the knife.
Fishing deep within the fissures.

Glottal emissions
stream and curve
into a common air.
Mutter matters.
Signifiers.
Ferret out the *there*.

Fragments prostrate, dark on light.
Blade trims and sets the accents right.
Scraps and sketches shifting yet.
(Suggestions of a sapphire jet?)

Snaking through the transport
mouths open at the gate.
Gather up the snake skin,
stitch and baste the speech and babble.

Spills a coiled mass
as severed subject drains
out of the bedded length.
Alluvial terrain.

In the zones of contact:
style is subject to the steel.
Traffic is directed
as blade runs from reel to reel.

Directed, bisected,
detoured, rejected.
Flesh, curved, grasps to slash.
Cutlery? Connected.
(Sutures, save the snakes.)

Sound limps or stamps into the room
of bright hotels, or else in tombs
cut tongues floor-fall. Enjoin the sins
and fade to sound of violins.

Exiled from gregarity,
united in congruity,
edits bite at the site,
which affirms sound practice.

However soon, however late,
Guillotines do punctuate.

From One Head to Another

René Farabet

What Ulysses seems to tell us, his body arched against the ship's mast, is that listening involves a sort of rapture, a transporting movement, a movement of surrender and of desire. Perhaps his ruse was to have guessed that beyond the tympanum is a minuscule vibrating bone, the ossicle, that allows the inner ear to unroll, so to speak, like a camera's lens, creating a zoom effect. And it is thus that a hero in chains could believe he would seize the sirens' song without risking the debris of his bones turning white on their shore.[1]

Such ecstasy, alas, was forbidden to his companions: their ears had been sealed with wax. Much later on, in order to escape different sirens, the city man will pull his hat down on his head: felt, then, replaces honey. Then headphones will come. The urban noises are lost in fabric or bounce back from plastic shells. Look closely: under this mass of animal hairs, like under the shiny plastic, the head nods forward with the force of an internal turbulence; the eyes are vacant; the isolated man no longer greets anybody; he becomes deaf to others. Nothing remains in him but this "interior radiophony" that Barthes spoke of, which transmits incessantly in order to fill up the void, to stir up gurgling whirlpools that instantly drown out all outside cries. There is a perpetual warning signal alight atop Western man's forehead, while under his hair ensues a warlike racket which is in fact an activity of interference, of *trompe-l'oreille*. Who knows if at a certain moment Ulysses himself, the old pirate, hadn't caused some sort of interference, short-circuiting the melodious voices in his head in order not to be trapped?

But deep down, must one still feign to believe in this hypnotic abduction, in this bewitching force capable of detaching a man from himself, of binding all his senses and delivering him entirely to the other, an ecstatic and voluptuously passive victim, exalted with joy? For even at the peak of his desire to listen, Ulysses still belongs to his foamy context. No doubt the concert was disturbed by the panting of the crew, the batting of the oars, the lapping of the waves. And if the eye looked up, it would be lost in the fantastic forms of the clouds blown by the wind. Fleeting forms, the curve of a neck, a sheaf of curly hair... We know well, and Cage has laughingly reminded us of it, that "pure" sound does not exist; there are nothing but listening-situations. Blindfolds and gags are equally inefficient; in every soundproof room, there will always be a book with an inviting surface, or a familiar object to which memories cling. And no padding will silence the rush of blood, nor the more deafening train of thoughts. The auditory field is a field of wandering. The old husband of former times would forbid his new wife to approach the window, but interior life, in fact, is a succession of defenestrations.

What, then, is listening?

You have decided to listen attentively. You turn on the radio. It spits out a waiting, crackling sound. A voice, for example, comes through to you as though imprisoned in the box: it's a prisoner's song. The song, yes, but not the prisoner—that which elsewhere one would call a "partial object." And yet it seems so close—you are neighbors—and perhaps you would like to believe that it is to you alone that one is speaking. But you know well there is no face-to-face. That word passes over the shoulder, as though launched by a distracted discus thrower. It's a game of profiles, like that of the silhouettes in a shadow theatre. The rustling sound that comes to you is but a faint, hazy trace, the residue of another celebration, an absent celebration that has already taken place (even though it is felt to be synchronous). You are not, then, behind closed doors, in police quarters. Allow yourself to be cast, like a yo-yo, upon the sound's trajectory, between the faraway source and the impact of the sound resounding in your ear. This is not a linear track. The sound does not project like an arrow. It radiates, it explodes. And in fact, you already are in an imaginary space, at the heart of a dreamy, utopian activity. Through what you hear, let yourself drift upon an unstable raft, surrendering to shifts in space and time, to plural presences, a multiplicity of "listening points."

If the eye remained fixed before the image, as though dumbfounded, it would capture nothing but clichés. It must blink in order to sharpen its gaze, its perceptive vivacity. As for the ear, it too vacillates between attention and absence. The sound itself carries the mark of these lapses of attention, these comings and goings of consciousness to which the first listener, the author, is subjected (and so it is of him that we have been speaking so far). Whichever sirens he might convoke, his murmur drowns their charmed song. He knows it is deceitful like a momentary calm. And should he lose himself amidst the delicious airs, he holds in his hand a blind man's cane, a balancing pole that allows us to listen to the entire world without succumbing to it. Every composition is thus a mix of a charmed voice and a coughing fit. It is true that once it is completed, worked out to its "utmost intensity," as Bresson would say, meaning to the utmost force, the murmuring sculpture might appear at times as fatal as a woman in a fish's body, its tightly woven texture seemingly impossible to tear. But there will always be someone to tatter the drapery, tearing it to pieces, to stones—decapitated bodies, torn-off limbs—returning it to the stone yard, to this grand quarry that is the world. Every *au(di)teur,* in a sense, is a vandal. Yet from the fragmentary perception, a global image—a "listening proof"—will at last be extracted. Listening is advancing upon a gradually burning earth, reading a charred score. It is the activity of a watchman, not that of a sleepwalker.

Kafka imagined that the sirens, their mouths open, had suddenly stopped singing, confident in the power of their silence, or rather in the lack of their song, to intensify desire. After all, Munch knew how to make us hear this silent cry. It is on such a silent ground that the radio man works. And the listener must not disregard him even if, like Ulysses, he feigns ignorance. The things around us sound on; sound continues to capture us. It is up to us to find breaches, points of evasion, to work upon the echo, the trail, the fading, the suspense, the intense slowness of silences... It is there, without doubt, that man is at his most imaginative, but also the most vulnerable, because he is closest to himself. Yet in the night of sound, the voyager's ship stands a stronger chance of crashing against the rocks, terminating his adventure.

Was not the sirens' island, after all, but a mirage, a virtual image?

—*translated by Talya Halkin*

Note

1. "D'une tête à l'autre" originally appeared in a dossier edited by Allen S. Weiss entitled *La création radiophonique,* in *Java* 12 (Winter 1994/95).

Radiophonic Ontologies and the Avantgarde

Joe Milutis

Introduction

For Artaud, "an expression does not have the same value twice, does not live two lives; [...] all words, once spoken, are dead" ([1938] 1958:75), and this unwholesome aspect of language, when coupled with the incantatory and vibratory properties of radio, propels what Allen Weiss, in an essay on the work of Gregory Whitehead, describes as the project of radio art: "Radiophonic art is guided by the serendipity of a fata morgana, the bewildering, aleatory process of recuperating and rechanneling the lost voice" (1995:79). That is, *in the one ear,* we have the poststructuralist scenario (inaugurated by the scenographemes of Artaud), in which meaning progresses noisily, without stable referent, as one word cannot double or replicate another in intent, force, meaning, or effect. Yet, *in the other ear,* in its struggle to rechannel loss, the art of radiophony attempts to circuit language back to some original, predictable, even replicable source in the living human body, even though this circuit is formed by chance operations in an illusory referential system.

In Whitehead's *Dead Letters* (1994), postal clerks in the dead letter office become an apt subject for the radio artist, as they echo this serendipitous rechanneling of loss in their quixotic attempts to resuscitate nixies and redirect them towards their intended, living addressees. The art of radio, like Luigi Russolo's *Art of Noise,* is invested in "choosing, coordinating, and dominating all noises, [...] enriching mankind with a new unsuspected voluptuousness" ([1911] 1986:171), paradoxically recuperating the referent without mimetically reproducing "life." Reproduced mechanically or mimetically, life is actually death, a paradox that is most obvious in the "live" aesthetics of broadcast media:

> [R]adio is actually at its most lively when most dead. Since the living cast themselves out through the articulated corpses of advanced telecommunications equipment, the whole idea of "live" radio is nothing more than a sensory illusion. [...] The more dead the transmission, the more "alive" the acoustic sensation; the more alive the sensation, the more "dead" the source body has become. (Whitehead 1991:87)

The sensations of avantgarde radiophonic art, mediated by articulated corpses, are counter-articulations of a life-force behind the death masks of electronic reproduction. If the (electronic) reproduction of life is actually death, then radiophonic sensations are only communicable by an antireproduction based on chance, conjuring up the "body electric." For example, the work of John Cage utilizes aleatory devices in order to inhabit the radiophonic universe without reproducing it in art, pointing in a Zen manner to what cannot be an object of the pointing—the invisible noise of electronic culture, source and substance of radiophonic ontologies. In a *Radio Happening* with Morton Feldman, Cage says,

> But all that radio is, Morty, is making available to your ears what was already in the air and available to your ears, but you couldn't hear it. In other words, all it is is making audible something which you're already in. You are bathed in radio waves. ([1966–1967] 1995:256)

What the work of much radio art reveals is the struggle to reveal the already there. Many times the desire to reveal the invisible, immaterial, and essentially unrevealable substance of radio (beyond the actual institutions and technology of radio), takes the form of a struggle to manifest the radiophonic as reality itself, part of our basic make-up. Even though radio's ethereal and vaguely metaphysical aspects might seemingly relate it more to superstructure and false ideology than to true matter, radio is a thing of matter, even if it is a matter that struggles to be known, always to be suppressed. While, in the Cage worldview, a rock is a radio radiating molecular waves, *radiation*—in the post-Enlightenment, post-Chernobyl, and post-ozone world—is that unwholesome glow from which we protect ourselves with the second skins of sunblock, safety procedures, and cynicism. Avantgarde radio art attempts to create a sonic bridge through the inscrutability of dead signs (a derma protecting us from the radiation of *the thing itself*) to the real of radio, even though it is fully aware of the impossibility of recovering the real through practices of representation. I will touch upon these particularities of the radiophonic and the avantgarde practice of radio art before discussing the radiophonic aspects of specific experimental dramatists and performers who were obsessed by the simultaneous promises and difficulties of producing an art uniquely "for radio." Along the way, we might find that the term "art radio" is oxymoronic, since it elides the incompatibles of form (art) and noise (radio). The Futurists, Brecht, Artaud, Beckett, and, to some extent, members of the other avantgarde movements (Dada, Expressionism) meditated on the external manifestations of this interiorizing technology, a technology that creates a highly contested space where space is contested, and that provides a context in which stages and scripts may liberate themselves from context itself.

Cage's aforementioned innocence about electronic culture (the dominant paradigm of which, I would argue, is radio, not TV) in the *Radio Happening* is in counterpoint to Feldman's initial cynicism towards Cage's happy ebullience: "I can't conceive of some brat turning on a transistor radio in my face and saying, 'Ah! The environment!'" ([1966–1967] 1995:256). There is a sense that radio reality is not just "there," but that it intrudes and colonizes, its "imaginary landscapes" making impossible "imagined communities," thought, or solitude in an electric company-sponsored disruption. Radio art bridges this ambivalence between celebration (the Cage standpoint)[1] and cynicism (the Feldman standpoint), knowing full well that the risk of life *between* these two points, in the electronic chaos, challenges the importance of artistic personality and aesthetic judgment. (After all, the cybernetic scenario is the locus of authorial death.) More importantly, perhaps, the space between possible judgments *of* electricity is the moment when electricity judges, manipulates, and "bathes" *you*, heralding the loss of coherent bodily sense. Artaud in particular, in giving his body up to electromagnetic waves, became a body without organs. Radio art, as in Marinetti's *Variety Theatre* manifesto, encourages this *fisicofollia* or "body madness" ([1913] 1986:183) of a body under the electrical regime, where inquiries into truth receive static back, unlike the regime of the coherent organism that "knows" itself only because of a highly disciplined closed circuit. Avital Ronell says of a schizophrenic's radiophonic experience:

> Her "word salad" seems to be the result of a recording, registering a number of quasi-autonomous partial systems striving to give simulcast expression to themselves out of the same mouth. [... T]here is a lack of overall ontological boundary. (1989:147)

Radio's most fundamental, ontological feature is precisely this ability to break down ontological borders, a process which is very similar to certain forms of psychosis. There are two dominant forms of psychological disorder that the radio environment mimics and enhances. In the first, the radiophonic universe takes the voice away from the body, stealing words—as in Artaud's paranoid scenario—and transmitting them everywhere. This ability of the radiophonic to steal words and thoughts is evident even in the most wholesome productions of Golden Age radio, all of which, by convention (especially the "thrillers"), have the interior thoughts of the character closest to the microphone "revealed" to the mass audience, so that, in the delirium of reception, the listener's thoughts are replaced by the protagonist's in an identification structure unique to radio. This psycho-narrational aspect of Golden Age radio crosses over into the *noir* productions of the time, in which the interiority of the voice-over, emerging from a wounded or pursued body, "implies linguistic constraint and physical confinement—confinement to the body, to claustral spaces, and to inner narratives" (Silverman 1988:45). This claustral point-of-view, when not subject to the limiting image (as in the noir film) gives the listener no basis for discerning whether what is narrated is the product of his or her own interior delusion. Thus, the paranoia of stolen, surveilled thoughts is compounded by the paranoiac anxiety that the thoughts returned in exchange for the stolen ones are all lies (a repressed fear that is manipulated in Welles's *War of the Worlds* broadcast [1938]).

Secondly, in a disruption of the coherent, yet generally unhealthy, interiors of Golden Age radio and noir film, radio loads more voices into the head than the body can withstand—the "schizo*phonic*" condition that Whitehead maps. Avantgarde radio exploits the schizophonic, overcrowding the interior space of radio reception with many voices and sounds, disrupting traditional visions of what the tape, music, or the interpretive apparatus behind the ear can withstand. In Whitehead's *Pressures of the Unspeakable* (1992a), the nervous system of Sydney—a city reconfigured as a schizophonic body—is mapped radiophonically by the recording of inhabitants' screams on a 24-hour "screamline" (Whitehead 1992b:115). The interpretation of these various screams—some of which seem to overload the recording equipment—is performed by Whitehead himself as "Dr. Scream." As ironically calm doctor and narrator, his clarity belies that he really has no control in this dispersed nervous topography: "What is certain is that this 'nervous system' is simultaneously that of Sydney *and* of Whitehead *and* of radio circuitry—all of which coalesce into a possible alter-ego for the moments of our most severe nervous tension" (Weiss 1995:83). Whitehead's radio art is based on a "*principia schizophonica*" which Weiss argues is part of the ontological structure of radio: "In radio, not only is the voice separated from the body, and not only does it return to the speaker as a disembodied presence—it is, furthermore, thrust into the public arena to mix its sonic destiny with that of other voices" (1995:79). Because of this paranoid-schizophrenic stereophony, even though radio is omnipresent, the radiophonic eludes psychic as well as institutional organization. To rephrase the evangelical aphorism, *Radio is Love.*

Thus when the radio body has not entirely disappeared, as on the Futurist stage ("a colorless electromechanical architecture, powerfully vitalized by chromatic emanation from a luminous source" [Prampolini (1915) 1986:204-05]), it is presented as a mad body in historical radio art (Futurists, Artaud, Beckett), a body beyond the modes of reason that reason has presented, a body like Cage's prepared pianos in which the "natural" vibrations are deflected by "technological" intrusions, which the Futurists called *excitations.* No longer do nerves excite other nerves in a narcissistic closed circuit. Rather, from the Futurists on, the body's signals are deflected and cybernetically connected up with signals that have more intelligence, freedom, and futurity than common-sense language. These signals are sometimes literally digital, as in Giacomo Balla's pieces in which numbers are recited as part of the glossolalia ([1916] 1986:232-33). The body vibrates erotically through contact with out-of-body signals that deconstruct, as Marinetti claims, traditional psychology.[2]

This body-madness, if survived, promises a transformation through decomposition. Bodies become "exultant, luminous corporalities" (Prampolini [1915] 1986:205) in the dark of radio's theatre. Formerly constrained by provincial intelligence (the source of irritation especially for the Italian Futurists), the body realizes fantastic possibilities. Fortunato Depero's theatre calls for "[d]ecompositions of the figure and the deformation of it, even until its absolute transformation; e.g., a dancing ballerina who continually accelerates, transforming herself into a floral vortex, etc." ([1916] 1986:207). If *stunad* (from *stonato*) is a damning epithet in Italian meaning not only "out-of-tune," but also "a little crazy, a little stupid," the Futurists and other radio artists risk cultural damnation by intentionally voyaging out-of-tune. Perhaps, more accurately, they voyage out-of-form, risking stupidity, or out-of-body, madness, in order to rechannel and repackage sensation, noise, and communication—momentarily spanning a bridge between technological and biological noise, going beyond language to the blissful vibrations of the thing itself.

Since no one concept of "out-of-form" can be correct without instituting another monolithic concept of form, sanity, or reason, the body of radio art work is dispersed and undisciplined, posing difficulties for the historicizing of radio art within sound history; radio is supposedly perceived only in the interior space of the mind, an intimate space incommensurable to historiography. The attempt to organize radiophonic noise on a wide scale—no less the range of concerns of this essay—has always met its challenge in this intensely personal space (akin to the presocial or maternal) where radio is received. From Marinetti's "pure organization of radiophonic sensations" (Zurbrugg 1981:54) to the creation of profits out of ether by cyber-industries and Wall Street financiers, from the Bible's erasure of the Big Bang to Fanon's description of radio backfiring on Algeria's colonizers—one can see how "nationalistic" projects to organize and use radiophonic chaos are always undercut by the crashes, the revolutions, the noise, and the nonsense of a radio-engendered universe. Even though large, state-financed broadcasting has traditionally used radio to construct a national voice, radio art is incommensurable to this project of unification and *whole*-someness. It has thrived in pirate radio, community radio, anti-gallery gallery installations, tape culture, avantgarde film and performance—illuminating the solitude of production and consumption of an unprofitable art which does not attempt to conquer space and time. In fact, contemporary radio art, even more so than the radio and sound art of the '50s and '60s, is engaged in the act of *hysteron proteron,* turning back the technological clock in the face of technological hype, reinjecting the primal into the postmodern, making the future strange by the avantgarde use of an "obsolete" technology.[3]

This dialectic between future and past has always been an aspect of avantgarde art: "[I]t should not be forgotten that both the Modernist and Post-Modern avant-gardes evince a 'zero' phase, in which aspirations to what Gysin terms 'machine poetry' are counterbalanced by 'primitive' alternatives, deriving inspiration from the distant past" (Zurbrugg 1981:54). However, as never before, radio, once the sign of future aspirations, now signifies the past quite efficiently. Even though William Burroughs, cut-up tape artist, has made it into some now infamous Nike ads, and Joe Frank, late-night radio monologist and experimental radio dramatist, hawks Zima—seemingly unifying their vocal personalities with a thousand points of light—these moments are rare, targeting a small audience and by no means heralding the reinvigoration of radio art in U.S. television culture. Radio's "Golden Age"—the only area of interest to the few publicly accessible radio archives in America—is over. However, radio—for its own avantgarde and for outsiders—is the future and the past, coursing through the century, creating and destroying, an immaterial primal matter so unstable and creative as to make apocalypse obsolete and beginnings interminable. Radio is the suppressed double of our visually material universe.

Bridging the Gap

The Proles of the Synapse

> *Radiophonic space defines a nobody synapse between (at least) two nervous systems. Jumping the gap requires a high voltage jolt that permits the electronic release of the voice, allowing each utterance to vibrate with all others, parole in libertà. Or, as fully autonomous radiobodies are shocked out of their skins, they can finally come into their own.*
>
> —*Gregory Whitehead (1991:85)*

A dispersed nervous system, in constant crisis, evident in radio works like *Pressures of the Unspeakable* or Artaud's *To Have Done with the Judgment of God* (1947), is the already operative precondition for dissolving the distance between word and thing, theatre and life, facilitating either the revolutionary leap into new perceptual and productive relations or the descent into madness. The synapses firing, "there will be neither respite or vacancy in the spectator's mind or sensibility. That is, between life and the theater there will be no distinct division, but instead a continuity" (Artaud [1938] 1958:126). And this continuity, created in the collapse of the boundary between public representation and private reception, uniting real and illusory, is described by Artaud in ways suggestive of the radiophonic flux beyond the image of "life" reproduced in the traditional, psychological theatre: "Furthermore, when we speak of the word 'life,' it must be understood we are not referring to life as we know it from its surface of fact, but to that fragile, fluctuating center which forms never reach" (13). Artaud's radiophonic experimentation espouses a dark Platonism in which formal representation never reaches the realm from which representation emerges. Most likely this realm is the body, the dark reality to which the radiophonic accedes. The paradoxical anti-formalism of radio art nevertheless attempts to reveal this suppressed underside of theatrical representation and of representation in general.

In *The Theater and Its Double*, Artaud introduces as a method of spanning the gap between sign and signified a poetics based not in representation, but in the unsettling notion of the Double:

> [T]he theater must also be considered as the Double, not of this direct, everyday reality of which it is gradually being reduced to a mere inert replica—as empty as it is sugar coated—but of another archetypal and dangerous reality, a reality of which principles, like dolphins, once they have shown their heads, hurry to dive back into the obscurity of the deep.
>
> For this reality is not human but inhuman, and man with his customs and his character counts for very little in it. Perhaps even man's head would not be left to him if he were to confide himself to this reality [...]. ([1938] 1958:49)

This reality very much resembles the cybernetic, radiophonic, and fluid universe—a dangerous universe for Artaud, who attempted to counteract the effects of electroshock therapy with his own shocks to the radio system in the scatological and eventually suppressed *To Have Done with the Judgment of God* (see Weiss 1992:271). Spanning the gap between signifier and signified, disrupting localized signifiers of madness and displacing them, hurling them free of the body into the electronic and disembodied politic, radiophonic art such as this continues a fantasy dreamt up by the Futurists, the fantasy of *parole in libertà* (words in freedom).

"[W]ords-in-freedom [...] smash the boundaries of literature as they march toward painting, music, noise-art, and *throw a marvelous bridge between the word and the real object*" (Marinetti [1916] 1986:214, italics added). The freedom that the Futurists sought is perhaps the freedom of the word to merge with the real—an impossibility for those who have Lacanian turntables. This bridge has been out, deconstructed, as it were. Only the words in freedom—here represented as proletarians of this futile endeavor (smashing, marching, building)—remain. Bodies, translated into words in freedom and dis*organ*ized, rechanneling libidinal transportation into a new technological reality, smash the traditional boundaries of illusion. This bridge, in Dadaist Tristan Tzara's terms, makes seemingly parallel lines meet by utilizing "the supreme *radiations* of an absolute art" ([1916] 1987:47, italics added) and thereafter making possible "the elegant and unprejudiced leap from one harmony to another sphere; the trajectory of a word, a cry, thrown in the air like an acoustic disc" (51). In Artaud, a literal painting of a bridge represents for him another, internal bridge that blurs the concrete and the metaphysical:

> [W]itness for example the bridge as high as an eight-story house standing out against the sea, across which people are filing, one after another, like Ideas in Plato's cave. [... T]heir poetic grandeur, their concrete efficacy upon us, is a result of their being metaphysical; their spiritual profundity is inseparable from the formal and exterior harmony of the picture. ([1938] 1958:36)

Surface harmony and spiritual depth are linked in the moment of a dematerialization that facilitates a dangerous perceptual span between subject and object. The artist's delirium generated out of this perceptual connection is given an elusive but nonarbitrary structure (why eight stories?), momentarily containing the delirium in a concrete image in order to communicate the metaphysical. This bridge, a strangely visual and material image, is perhaps built at the expense of total (and destructive) *jouissance*. The function of its materiality is to present a *new relation*, rather than a non-relation, between signifier and signified. There is, in effect, a politics to delirium.

In his introduction to *Phantasmic Radio*, Allen Weiss introduces the phenomenology of radiophonics, not only as the future of radio, but as an addition to contemporary theoretical paradigms, an addition which rethinks the past and restructures the future in terms of radio. I find his explication of a new bridge between signifier and signified compelling:

> [R]adiophony transforms the very nature of the relation between signifier and signified, and [...] the practice of montage established the key modernist paradigm of consciousness. This task is informed by the *motivated, non-arbitrary* relationships between signifier and signified (S/s), where the mediating term is not the slash that delineates the topography of the unconscious (/), but rather the variegated, fragile, unrepresentable flesh of the lived body. As such, this work participates in the linguistic and epistemological polemic at the center of continental philosophy—between phenomenological, structuralist, and poststructuralist hermeneutics—concerning the ontological status of body, voice, expression, and phantasms. [...] Between voice and wavelength, between body and electricity, the future of radio resounds. (1995:7–8)

This shift from *the unconscious* as the mediating term to *the body* is all important, although quite difficult to conceive.[4] In Lacan's scenario, what is signifiable submits to *extracorporeal* relations (the unconscious) in order to produce a signifier. These out-of-body relations determine the "it" that speaks through the subject, and thus we are always dealing with the Other when the "I" speaks. This problematic of language is the basis of the idea that pain cannot be communicated, since bodily sensation is radically subjective: the state of the body cannot be spoken through language without a misrepresentation or misrecognition. However if, as in radio, one considers the extracorporeal not as a superstructural presence but as *the very material of radiophonic corporealities*, then we have an entirely new paradigm to consider.

While in Lacan's scenario we are radios that speak the transmissions of an elusive source, in this "newer" radiophonic scenario, the body is source, substance, and medium of radio. Not only is the whole body considered receptive to the whole gamut of signals and vibrations of the radiophonic universe, but the body also has an ability to transmit and record. The radio theatre is not just a place for the play of the disembodied image or imagination, covering up radio's perceived lack.

> While it has become a commonplace to talk about sound as the medium of the imagination (a gray area), the ear also opens a path for acoustic vibrations to travel through the spine and skeleton. Sound, then, is actually *a material for the whole body conducted through nerves and bones by way of a hole in the head.* (Whitehead 1991:85)

Here the lack, or the hole, speaks—the whole body is channeled through a hole in the head and through radio. Therefore, radio is not a medium discrete from the body. The radio artist is both producer and consumer, audience and performer, of his own electroacoustical soundings. It must be remembered that the structuring of everyday noises, including bodily sounds, as "music" (a Futurist practice reinaugurated by Cage) was in its time a controversial addition to the sensorium of reproduced sound. Furthermore, like the body artist (many of whom, including Vito Acconci, Dennis Oppenheim, and Terry Fox, engaged in sound art), the radio artist, by introducing the body, dematerializes the art object into the performing presence. Like body art, sound art, when it utilizes the clicks, the hums, and other extralinguistic bodily manifestations of the voice as its material, is transmitting, as if from the living to the dead, a "new *aesthetics of existence,* [...] seeking to suppress the aesthetic illusion, exceeding traditional aesthetic bounds and classifications in terms of dancing, theatre, or films, once again drawing closer to that heterogeneous totality of experience that we know from everyday life" (Gorsen 1984:141). The body is thus an integral part of the transmission/reception complex of radio art, even though common images of radio airwaves present an ethereal realm where signals play separately from the grounded body. For Whitehead, the ear is the bridge between the ethereal and the bodily, expanding the domain of radio's electronic play and transforming the body into a player. William Burroughs, performing a monolog as Mr. Martin, a U.S. citizen who has been sent up into space and who, upon his return, is mistaken for an outer space alien because of his newfound disdain for humanity, remarks, "Human activity is drearily predictable. It should now be obvious that what you considered a reality is the result of precisely predictable because pre-recorded human activity. Now, what can louse up a prerecorded biological recording?" (n.d.). Burroughs's cut-up method, like Whitehead's, redirects the flow of information by cutting into the recorded transmissions of the mass media with biological recordings. We hear his body, and his full-bodied voice, then, through the disembodied signals of the mass media. He deforms the consumption and reproduction of dead forms that compose "live" radio. Burroughs performs an antireproduction based on internally motivated chance operations (almost surgical—cuts without anesthesia) rather than external form. The body becomes a *radio system* (in the chaos theory sense of system) rather than a *radio set.* It is transmitter, receiver, and director in one.

The lines between production and consumption are broken down in this system, and a circuitry is set up so that what were once separate spheres continually modify each other. The real of radio is released, and pleasure is rechanneled as the body becomes part of a "bachelor machine," as in the anti-Oedipal scenario. In *Anti-Oedipus*, Deleuze and Guattari break down the stage of traditional psychoanalysis in a radiophonic manner, opposing their circuitry to a massified, standardizing discipline. They describe the underside of the productive universe with a metaphor of constant recording:

> For the real truth of the matter—the glaring, sober truth that resides in delirium—is that there is no such thing as relatively independent spheres or circuits [read: independent bodies, technologies]: production is immediately consumption and a recording process (*enregistrement*), without any sort of mediation, and the recording process and consumption directly determine production, though they do so within the production process itself. ([1977] 1983:4)

Even non-radio artists have taken up the metaphor of body as both performer and that entity which is sounded against (audience). The body, resonating between "I" and "Other," transmits its resonations in order to liberate the body from its Western Instruction Manual, but only at the risk of a "raving consciousness" (Kozloff 1975:32). (For whom is this consciousness raving? Where does the burden of this perception and interpretation lie?) For example:

> Joseph Beuys, lying face down for three hours in a Naples gallery, rubbed his oil-smeared hand over copper slabs until, as a writer has described it, "his body vibrated loaded with energy like a body charged with electric current." The most recurrent sentence is: "I am a transmitter. I emit." (Kozloff 1975:33)

This reorganization of the body not only as receiver and producer, but also as transmitter, carrier, and ultimately disrupter, highlights the "sober truth" in delirium rather than the pathology of delirium. Without the topology of the extracorporeal Other, which Artaud disdains, it is impossible to record, reproduce, and recognize the signs of psychosis except as a total condition—the truth of the body electric. Without the "it" speaking through man, "it is impossible even to register the structure of a symptom in the analytic sense of the term" (Lacan [1958] 1982:79). Notice Lacan's use of the word "register," which can imply the act of recording a tape. Artaud then, in his disdain, seems to be disrupting psychiatric symptomatology in his theory and radio work by disrupting the dictates of faithful recording. He attempts a return to the what-has-been in a "magic identification" ([1938] 1958:67) with an unrecorded past of communal wholeness. In the act of suspending our modern disbelief in the communal possibility of a dispersed stage, he displaces diagnosis onto the body politic, further reducing the identificatory structures of both the everyday and the psychoanalytic to noise: "WE KNOW IT IS WE WHO WERE SPEAKING" (67). Everybody risks psychosis, and the only way French radio could quash a postwar psychotic crisis was to contain the broadcast on the tape, deadening it in a magnetic crypt and not allowing its supernatural qualities to awaken the dead of the airwaves. Pathologizing the tape itself, and suppressing the necromancy of the text, French radio answers the unsettling question "Is it live, or is it Memorex?" by siding with the tape, in the hopes that the unwholesome utterances will not surpass the tape's dead materiality. Artaud's answer to the question, "Is it live, or is it mimesis?" might choose both, aware of the unsettling nature of the Double. Symptoms are a mimetic illusion that contains the living structure of a sickness as if on tape. The act of registration and the ideologies of tape repress the psychotic underside of postwar radio culture—a reality of fragmentation, shell shock, and exploded identities. The diagnosis is always another dead repetition, the living sickness beyond the reach of the speaking cure; the spoken, the enunciation of "it," masks the truth of a total delirium and derangement experienced everyday by vibrating bodies.

For Brecht, the revelation of the vibration between bodies—animal desires in the dark that conquer even the thick-skinned—is part of his interactive Marxist concept of theatre. Not only does he delineate the barriers between alienated characters in the hopes of vibrating them out of those barriers, but he also foregrounds the edges of theatrical illusion, the better to dissolve them as well—uniting audience and stage, and creating new relations. His dream of radio is one in which the audience both receives and transmits, bringing something new to every performance. Perhaps Brecht, more materialist than alchemist, has ambiguous feelings concerning the actual effectivity of a bridge between word and thing constructed outside the lights of theatre. I sense this hesitancy to embrace the Platonic cave of radio in a comment on actual reproduction, spoken by Garga of *In the Jungle of Cities*:

> Love, the warmth of bodies in contact, is the only mercy shown us in the darkness. But the only union is that of the organs, and it can't bridge over the cleavage made by speech. Yet they unite in order to produce beings to stand by them in their hopeless isolation. And the generations look coldly into each other's eyes. ([1927] 1971:157)

Perhaps the bridge between the spoken and the real that sound constructs is only done by a ruse "in the dark." Perhaps the utopian or dystopian radiophonic universe, if experienced, is only a momentary gratification, and will lose its transcendent power in the cold light of vision, an inevitable event in our psychic economy. The next section will deal almost exclusively with the shorter works of Beckett—works which, as will become evident toward the end of my argument, highlight the unavoidable dialectic between hearing and vision, even in works that are limited to the sonic realm. I will deal initially with the aspects of Beckett's drama that most successfully point to a buzz and hum behind the Word, a seething subsensory substance, and I will then consider how the economy of vision torments this substance into appearance.

Molecular Orality and the Vision of It

> *The original speech act begins to disintegrate as soon as it comes to grips with its schizophonic double.*
>
> —*Gregory Whitehead (1990:60)*

> *Do you find anything ... bizarre about my way of speaking? (pause.) I do not mean the voice. (pause.) No, I mean the words. (pause. More to herself.) I use none but the simplest words, I hope, and yet I sometimes find my way of speaking very ... bizarre. (pause.)*
>
> —*Samuel Beckett,* All That Fall *([1957] 1984:13)*

Beckett's plays are uniquely oral plays; if they do not explicitly engage with the radiophonic (for example *All That Fall, Embers* [1959], *Cascando* [1964]), they limit the multimedia possibilities of the traditional stage in order to direct the visual and aural attention of the audience to something like the radiophonic. Plays such as *Play* (1964), *Not I* (1973), and *That Time* (1976) make their protagonist the voice and their antagonist the body—paralyzed by age, pain, memory, or surrealistically incarcerated by such devices as the urns of *Play*. One of the voices in *That Time*, a play with many voices trapped in a single head, says, "no notion who it was saying what you were saying whose skull you were clapped up in whose moan" (1984:231). Radio's clichéd but celebrated "theatre of the mind" is transformed into a nightmare space of schizophrenia and melancholy, where one's most intimate thoughts can become alien entities when performed. Each speech act illuminates the drama of the cranial cavity's invasion by sense, an invasion which, as I have noted earlier, is the hallmark of the radiophonic.

One can say, *in light of* these oral dynamics, that Beckett's seemingly "shorter plays" are in actuality infinite plays, composed of hundreds of acts—speech acts—each with an infinite potential for interpretation. In contrast to traditional acts that mechanically push one another in fits and starts to the bitter end, Beckett's speech acts "act" as molecules do. The theatrical elements in Beckett's plays (for example, stage and body) are antagonized by their own brute materiality, seemingly doing nothing and going nowhere; however, these elements seethe with multiple acts of speech, a molecular orality. *Not I* stages the molecular orality of *decomposition*:

> so on ... so on it reasoned ... vain questionings ... and all dead still ... sweet silent as the grave ... when suddenly ... gradually ... she realiz— ... what? ... the buzzing? ... yes ... all dead still but for the buzzing ... when suddenly she realized ... words were— ... what? ... who? ... no! ... she! ... (*Pause and movement 2.*) ... realized ... words were coming ... imagine! ... words were coming ... (1984:218–19)

Words and flies buzz around the dead body. Vain questioning about the mystery of death hovers self-servingly over the corpse like the flies. All "nonperforming" bodies in Beckett, in their crepuscular or pathological, hypersedentary sentience, perform only molecularly and sonically, transmitting and receiving while fragmenting and decomposing, in the throes of radiation. Mrs. Rooney, in Beckett's most conventional radio drama, *All That Fall*, wails, "Oh to be in atoms!"—expressing a desire not only for death, but to be composed of small fragments free of the body and the sense of language. She perhaps wishes to transform her mythically huge and unmanageable body into a radio body. The simplest words become bizarre when free of the body, stripped of the illusion of the voice and sense, free to buzz in the radio airwaves with the flies. Mrs. Rooney's parodied desire for catholic transcendence of the human flesh generates a radio hallucination in which language breaks down into atomic particles.

For Beckett, then, the traditional fantasy of oral culture or radio culture[5] is perhaps an impossible dream of wholeness in a *particular* world. Rather than conjuring the song of a community (even though many of his radio works were famous for their productions on BBC—font of the British communal voice), his multiple-act plays and playlets present the utterly and irrevocably fragmented nature of speech. Spoken words cannot produce a cure for pain, even though some of his lines sound like parodies of any aspirin commercial: "all that pain as if ... never been" (1984:152). Many people have talked about radio's ability to form a coherent sound-image of the nation/everyman as a palliative for the ills of the body politic. This analgesic radio voice gives identity, direction, and coherence to the nation. In Beckett's plays, however, the (everyman) voice that is the inspiration for traditional fantasies of oral and commercial culture (and their combination in the advert: "Personally I always preferred Lipton's" [1984:154]) is replaced by a highly internalized, schizophonic voice in the head.

The voice in the head in Beckett sometimes lacks coherence to such an extent that it loses its moorings in the very head from which it originates. Does the voice belong to the head it inhabits, or is radio's "national" voice a colonizing one? In *That Time*, the multiple voices are incarcerated in a body that has somehow become alien to itself. "Could you ever say I to yourself in your life" (1984:230). There, in a *nut*, incidentally, is the postcolonial problematic: "no notion who it was saying what you were saying whose skull you were clapped up in whose moan" (1984:231). Question marks pleasantly disappear in these litanies, which are not meant to be spoken, yet are. Agrammatic thought, externalized, inexorably continues:

> not a thought in your head till hard to believe harder and harder to believe you ever told anyone you loved them or anyone you till just one of those things you kept making up to keep the void out just another of those old tales to keep the void from pouring in on top of you the shroud. (1984:230)

These plays have been called "skull-scapes": they dramatize the headache of having to constantly think ourselves, where each thought becomes an act or performance to keep out the radioactive void, even when acting, moving, or living is the least desired thing. By what mechanism is this internal thought brought to the surface in Beckett's plays? When we act alone, in our head, it is indeed an absurd drama, and not at all like the coherent, internal monologs of Golden Age and noir radio. And Beckett is maybe highlighting the sadistic nature of radio's intrusion that brings these absurdities to the surface, making the skull an unsafe place for the internal workings of the mind and imagination. "I can do nothing ... for anybody ... any more ... than God. So it must be something I have to say. How the mind works still!" says W1 of *Play* (1984:153), whom I characterize as playing a character, an everyman, only under duress. What is important is that this "radio nobody" is forced to be somebody *in the light of vision*. What "she" says can only be conceived as a masquerade of her internal thoughts, exemplifying the Artaudian belief that "the most commanding interpenetrations join sight to sound" ([1938]

1958:55). Encased in an urn, speech is her only possible action as actor, if she actually wishes to act. But she keeps on saying "Get off me! Get off me!"—ostensibly referring to the lights of the play, which elicit speech in *Play*, interrogatively, silently. The lights compel her to engage in the speech act, externalizing the internal, placing a gross beam on a dreamer whose inner lights, although dreamt, have already been extinguished, as in this Expressionist cry (here, from Kokoschka's *Sphinx and Strawman*) that claims that the stars of the soul pass only as Berkeley's tree falls:

> If I could only respond out of my loneliness to your secret confessions, oh, to be able to place a rainbow of reconciliation over shocked sexes, *(becoming hysterical)* my feelings are like so many falling stars, stars falling into the narrow fields of my soul to be extinguished—but the Word which reaches out far beyond me like a huge gesture means nothing to you. ([1907] 1986:33)

As W1 is compelled to make these speech gestures full of nothingness, the wholeness of her internal imagined self is fragmented into a multiplicity of acts which do not combine to tell any one truth, until the silence of death. The wholeness that the light presents is false (a discrete image), and the light also elicits sonic falseness, the lie of this externalization of the internal by speech. W1: "Is it that I do not tell the truth, is that it, that some day, somehow, I may tell the truth at last and then no more light at last, for the truth?" (Beckett 1984:153).

So, with the lights and language of the stage intersecting on these incarcerating urns, do we believe the "truth beauty, beauty truth" aphorism of the Grecian Urn poem, an aphorism which connects truth to vision? Or is truth beyond vision, in the molecular fragmentation that can be perceived behind the *trompe l'oeil* surfaces of abstract speech? In Beckett's *Film* it is noted that, "the protagonist is sundered into object (O) and eye (E), the former in flight, the latter in pursuit" ([1965] 1984:163). Even though this description might not include *Film* as a "radiophonic play"—it seems to be more about vision—Beckett's oral plays link the interpretative valuation of speech acts to the valuation of the object by the eye. Thus, in *Play*, while the register of action takes place entirely within speech, the speech is determined by the duration and locus of the light. "Being seen" (1984:157) becomes the same as being heard. Because of this dynamic, it may not be useful to distinguish vision and sound in Beckett (at least in *Play*), since, in a quantum world, both are products of particulate wave radiations. Arguably the perceptual apparatus of theatre, cinema, and television disciplines the traditional audience more to see than to hear, constructing differing levels of acculturated *perceivedness*; Beckett's plays, however, transform this discipline, and one cannot help hearing. But in all, despite the distinct and disciplined and sometimes deformed registers, the "agony [of the protagonist is] of perceivedness" (165), and the drama on Beckett's stage is a Houdini-like attempt to escape from the perceptual apparatus of the audience even while incontrovertibly *there*. Perhaps, then, we can replace the register of "vision" with "the perceived," and we will include sound on a different track of the same register rather than confuse sound, in a utopian leap, with that substance which falls out of relational structures of phenomenology. In the end, this utopian "substance" forms the metaphysical substance of the truth behind "shocked sexes," although we are left *speculating* as to whether it exists as the guiding force of the play. Is there anything in excess of perceivedness, in excess of the unreal structure of external values that creates the reality of the subject? If there is, all performative structures are in constant crisis, holding off the eruption of this subsensory matter. Beckett's plays—often called children of the nuclear age and, by association, our Emergency Broadcast Systems—manage these crises of immaterial power.

Play is a play about shocked sexes and the range of mastery—of this force behind appearances—that each character can manage. In general, different illusions of mastery of this force are available for different sexes, and this is the source of tension in *Play*. To what extent is the play "play" for each of the characters? M has enough mastery to call the past "just play": "I know now, all that was just ... play. And all this? When will all this—" (153). In the economy of perception, can we call M male and the W characters female? It would be interesting to see the choices directors make in this instance. Notwithstanding, W1 lacks the sort of mastery that M seems to exhibit. The lights force W1 to engage in "just ... play" (153) without advantage; the lights make torture for a soul that wishes to be quiet and to die. She feels merely played with rather than playing (157). The subject of *Play* is the manipulation of the play and here the manipulator is phallic. In *Play* the W characters are asked to compare happy memories (ostensibly regarding M) (154), so that, like the phallus, M is the standard for comparison and measurement, ratiocinating, the bar between two numbers in a fraction of desire. This mathematical relationality, coupled with the breakdown of discrete appearances onstage that the radiophonic aspects of this play enhance, points to the unperceived molecular substratum behind the realized hallucination of sex. At the same time, however, the lights peremptorily regiment reality as if in battle with this fragmentary and metaphysical substance. "Am I as much as ... being seen?" (157). M(an) can only be measured by his "being seen" not only as a stage actor, but by the two W(omen) in a sexual relation. He is both metaphorically and literally the dick, since his horror of the spiritual and Platonic implies that his relationships with those W(ithout) the dick were purely sexual. Even though sexual in nature, the play constantly bowdlerizes the explicitly sexual, since everything is limited to the seen, which in turn is regimented by the lighting. Even if the character that seems the most sexually comfortable in *Play*, W2, seems to experience the excess of *jouissance* in her "*peal of wild low laughter*" (157), this excess is cut off, measured by the time of the lights, turned into another value that the "mere eye" can discern. The "sense" that "being seen" creates is the source of all value in the theatre. And still, for Beckett, mere eye is not enough.

"Being seen" as the phallus is not the same as "being" or "having" the phallus. In Lacan and Freud, the distinction is made between the little boy, whose role is to *be* the phallus for the mother who, in an only deceptively coherent economy, desires to *have* the phallus. The subject's reality is created only through this unreal relation, the unreality of which is heightened by Beckett. Into this relationship, "appearance" (or masquerade) intervenes as a substitute for "having," and to mask the lack in "being." The ontology of the theatre has always been illusory—that is, it has always been about appearance (a word, by the way, with an inner "ear") and masquerade. It is never being or having which is played out, but appearing. And there is a sense that having and being are *never* played out, because it is only appearance that can extend out of the body as the body's mediating material; thus the theatrical metaphor has extended throughout even the most everyday activities. Yet the ontology of radio, as Herbert Blau has mentioned in conversation, is about the *shadow* of appearance—and "the Shadow knows." Whether radio is outside the theatre of the phallus, or whether, when we listen to radio, we merely "prick our ears" to the harmonic resonations of sex that prompt the phallus to the stage, is uncertain. Beckett's plays *of* uncertainty contain both theatrical and radiophonic ontologies and allow them to interpenetrate at the molecular and cultural levels of existence. Beckett's theatre of appearance stages disappearance even in the light of vision. And this disappearance is what Lacan calls *aphanasis*, or a fading, at the molecular level of language. For example, in *Play*, the equation between "being seen" and speaking (a paradoxical equation of the passive to the active) equates appearance with disappearance, vision's ruse with the lack that propels language into action. This radio-theatrical drama of Beckett dramatizes the speaking subject and compounds this drama dialectically with the economy of vision.

In the final fade-out, what the body is, what we hear in Barthes's "grain of the voice," is the not-body, the decomposition of the body. We were never "composed" except in some Platonic dream of hi-fi recording, or in the fantasy of digital remastering. Was there ever a mastering to begin with? What are we *masking* in the tape, except some backwards melody bringing us back to the source of all life—death? The radiophonic system—tape and razor, mike and mixer, transmitter and receiver—must always have an Emergency Broadcast System. This repressed double of the broadcast system, only returning with a vengeance in the threat of total destruction by catastrophic weather or the nuclear bomb, is contained in a test, only a test, a recorded tone of fixed duration. The composition of this tone is unsettling, and its repetition a denial of the constant reality of radiation and weathering which takes the body away, quanta by quanta, even as one hears the false subjunctive of "if this were an actual emergency." In a way, to "picture" this quantum reality of the body, one receives an image that resembles the image of consciousness, but also an image of war. Free of the body and emergency, both consciousness and words in freedom—which remain when the body and its voice are gone—give the taste of constant death. The voice, however, though constantly "signing-off" (the broadcast version of the swan song) and longing to merge with its metaphysical allies, articulates living presence on dead air. "Just one great squawk and then ... peace" (Beckett 1984:19).

Notes

1. Kathleen Woodward, in her analysis of the work of Cage, sees a fault in his uncritical embrace of "the electrical sublime," an idea that has been around since the 19th century and has only served to support the monopolies of the power and light companies (1980:189).
2. Whether these out-of-body signals are spiritual in nature was a source of contention for the Futurists. Radio's dangerous ability to vibrate the subject out of its borders is sometimes recognized as a spiritual quality of the radio. Futurists who were closer to Symbolism (like Balla) claimed that any dissolution of materiality, even if facilitated by technology, had to be spiritual in nature (Tuchman 1986:40). Later in his career, Marinetti repudiated Symbolism, constructing a more secular version of vibration, perhaps inspired by the very earthly vibrations of shell shock. Marinetti's version seems to have won out, if only because of the commodification and sexualization of vibration, repressing (or perhaps heightening) the transcendent qualities of exultant vibrations as they are incorporated in "Magic Fingers" beds and hand-held massagers. It is either Marinetti's dream or nightmare that late-night TV presents to our pre-REM retinas images of bikini-clad all-American girls shooting machine guns in slo-motion.
3. Perhaps the most well-known radio work that is popular for its use of the techniques of radio drama outside of their temporal context is Tom Lopez's *The Fourth Tower of Inverness* (1972), in which radio drama chestnuts are combined with the quirky mystico-political vibe of the early '70s. This shattered temporal soundscape is typified by the serial's magical "Lotus Jukebox," which determines the fate of the characters as it plays both '50s rock-and-roll and Zen koans.
4. In earlier formations of contemporary radiophonic art, the supple topographies of the body are elided with those of the unconscious, as when Gregory Whitehead, in a 1989 article, remarks, "writing radio puts into relief the supple contours of the human unconscious" (1989:11).
5. Whitehead describes this fantasy well:

> Every now and again, the quaint idea of radio as a kind of Talking Drum for the Global Village comes around for one more spin. In this romantic scenario, radio art is cast as an electronic echo of oral culture, harkening back to ancient storytellers spinning yarns in front of village fires. The idea has a seductive ring to it [... yet *m*]*ost* forgotten are the lethal wires that still heat up from inside out, wires that connect radio with warfare, brain damage, rattles from the necropolis. When I turn my radio on, I hear a whole chorus of death rattles [...]. (1991:88–89)

References

Artaud, Antonin

1958 [1938] *The Theater and Its Double.* Translated by Mary Caroline Richards. New York: Grove Press.

1992 [1947] *To Have Done with the Judgment of God.* Translated by Clayton Eshelman. In *Wireless Imagination: Sound, Radio and the Avant-Garde,* edited by Douglas Kahn and Gregory Whitehead, 309–29. Cambridge, MA: MIT Press.

Balla, Giacomo

1986 [1916] *Disconnected States of Mind* and *To Understand Weeping.* In *Futurist Performance,* edited by Michael Kirby and Victoria Nes Kirby, 232–33. New York: PAJ Publications.

Beckett, Samuel

1984 *Collected Shorter Plays.* New York: Grove Press.

Brecht, Bertolt

1971 *Collected Plays,* vol. 1. Edited by Ralph Manheim and John Willett. New York: Vintage.

Burroughs, William S.

n.d. *William S. Burroughs/Breakthrough in Grey Room.* Sub Rosa Records (Sub 33005-08).

Cage, John, and Morton Feldman

1995 [1966–1967] *Radio Happenings: Recorded at WBAI, NYC 7/9/66–1/16/67.* In *Exact Change Yearbook No. 1,* edited by Peter Gizzi, 251–70. Boston: Exact Change.

Deleuze, Gilles, and Felix Guattari

1983 [1977] *Anti-Oedipus: Capitalism and Schizophrenia.* Translated by Robert Hurley, Mark Seem, and Helen R. Lane. Minneapolis: University of Minnesota Press.

Depero, Fortunato

1986 [ca.1916] *Notes on the Theatre.* In *Futurist Performance,* edited by Michael Kirby and Victoria Nes Kirby, 207–10. New York: PAJ Publications.

Gorsen, Peter

1984 "The Return of Existentialism in Performance Art." In *The Art of Performance,* edited by Gregory Battcock and Robert Nickas, 135–41. New York: E.P. Dutton.

Kokoschka, Oskar

1986 [1907] *Sphinx and Strawman.* Translated by Victor H. Meisel. In *Expressionist Texts,* edited by Mel Gordon, 27–37. New York: PAJ Publications.

Kozloff, Max

1975 "Pygmalion Reversed." *ArtForum* 14, 3:30–37.

Lacan, Jacques

1982 [1958] "The Meaning of the Phallus." In *Feminine Sexuality: Jacques Lacan and the École Freudienne,* edited by Juliet Mitchell and Jacqueline Rose, translated by Juliet Mitchell, 74–85. London: Macmillan.

Marinetti, F.T.

1986 [1913] *The Variety Theatre.* In *Futurist Performance,* edited by Michael Kirby and Victoria Nes Kirby, 179–89. New York: PAJ Publications.

Marinetti, F.T., et al.
1986 [1916] *The Futurist Cinema*. In *Futurist Performance*, edited by Michael Kirby and Victoria Nes Kirby, 212–17. New York: PAJ Publications.

Prampolini, Enrico
1986 [1915] *Futurist Scenography*. In *Futurist Performance*, edited by Michael Kirby and Victoria Nes Kirby, 203–06. New York: PAJ Publications.

Ronell, Avital
1989 *The Telephone Book: Technology, Schizophrenia, and Electric Speech*. Lincoln: University of Nebraska Press.

Russolo, Luigi
1986 [1911] *The Art of Noise*. In *Futurist Performance*, edited by Michael Kirby and Victoria Nes Kirby, 166–74. New York: PAJ Publications.

Silverman, Kaja
1988 *The Acoustic Mirror: The Female Voice in Psychoanalysis and Cinema*. Bloomington: Indiana University Press.

Tzara, Tristan
1987 [ca. 1916] *Dada Manifesto*. Translated by Barbara Wright. In *Dada Performance*, edited by Mel Gordon, 45–52. New York: PAJ Publications.

Tuchman, Maurice
1986 "Hidden Meanings in Abstact Art." In *The Spiritual in Art: Abstract Painting 1890–1985*, edited by Edward Weisberger, 17–61. New York: Los Angeles County Museum of Art and Abbeville Press.

Weiss, Allen S.
1992 "Radio, Death, and the Devil: Artaud's *Pour en finir avec le jugement de Dieu*." In *Wireless Imagination: Sound, Radio and the Avant-Garde*, edited by Douglas Kahn and Gregory Whitehead, 269–308. Cambridge, MA: MIT Press.
1995 *Phantasmic Radio*. Durham, NC: Duke University Press.

Whitehead, Gregory
1989 "Who's There? Notes on the Materiality of Radio." *Art & Text* 31:10–13.
1990 "*Principia Schizophonica*: On Noise, Gas, and the Broadcast Disembody." *Art and Text* 37:60–62.
1991 "Holes in the Head: A Theatre for Radio Operations." *Performing Arts Journal* 13, 3:85–91.
1992a *Pressures of the Unspeakable*. On *Transmissions from Broadcast Artists*. Audiotape. Nonsequitor Foundation.
1992b "*Pressures of the Unspeakable*: A Nervous System for the City of Sydney." *Continuum* 6, 1. Edited by Toby Miller, 102–11.
1994 *Dead Letters*. Audiotape. Art Ear, San Francisco.

Woodward, Kathleen
1980 "Art and Technics: John Cage, Electronics, and World Improvement." In *Myths of Information: Technology and Postindustrial Culture*, edited by Kathleen Woodward. Madison: Coda Press.

Zurbrugg, Nicholas
1981 "Beyond Beckett: Reckless Writing and the Concept of the Avant-Garde within Post-Modern Literature." *Yearbook of Comparative and General Literature* 30:37–56.

Three Receivers

Douglas Kahn

Why would William Burroughs leave "three off-tuned radios blaring static" in his room in Tangier (Leary 1983:95)? Was he waiting for code, for voices, for an ethereal or chthonic broadcast? Cocteau's Orpheus tuned his car radio to pick up the latest from the underground. Leonardo heard voices on high in church bells, Joan of Arc, angels. Kerouac took dictation of his dialog with the waves and water of Big Sur. Dalí looked to seaside rocks of Cadaqués to interpolate signs from noise; Artaud, upon the land of the Tarahumara; and Ernst at the scratches, pits, and grain of floors and other surfaces. Did Burroughs hope to transcribe what the white noise said, to log its wisdom into what had captivated him for so long: the science—or the pseudoscience—of *fact*? In collaboration with Brion Gysin, Ian Sommerville, and others he had in fact carried out experiments using tape recorders, many of which incorporated radio sound and static. At times, time and its voices would leak through: one experiment announced the presidential foibles of Watergate a decade before they happened. We don't have to take his word for it. Whereas we have to take the word of Artaud or Kerouac for what they beheld, Burroughs appealed to phonographic repetition first for simple consensus and ultimately for clinical validation.

> There are then many ways of producing words and voices on tape that did not get there by the usual recording procedure, words and voices that are quite definitely and clearly recognizable by a consensus of listeners. I have gotten words and voices from barking dogs. No doubt one could do much better with dolphins. And words will emerge from recordings of dripping faucets. In fact, almost any sound that is not too uniform may produce words. (1985a:54)

Perhaps the sibilants and fricatives of radio static were too uniform to say too much to many others besides himself.

Dinbetween Stations

It was obvious Burroughs was a writer, for accompanying the radios was a desk cluttered with papers. Could the radios have been fulfilling a mundane requirement by supplying the room with a surrogate café raucousness? Walter Benjamin recommends that writers at certain phases within the production of a work seek out complex sounds:

> accompaniment by an étude or a cacophony of voices can become as significant for work as the perceptible silence of the night. If the latter

> sharpens the inner ear, the former acts as touchstone for a diction ample enough to bury even the most wayward thought. (1979:65)

The dish and din of café clamor can likewise soothe the conflict within the very act of writing—the gregarious motive of communication versus the solitude of its execution (Burroughs was known to talk ears off)—by providing a chatty noise within which a collectively discursive interlocutor can be divined. Noise also models the supple field of exchange between inner speech/sounds and those of the world and, thus, can situate the writer in this tender fray. It is commonplace, for example, for even the most dedicated musical aesthete to listen at times more concertedly to the psyche than to the concert, oscillating between stage and seat, constantly interrupting or melding in a mix that is, ironically, the means through which an idea of unity is negotiated. In the café where the sound is not the object of thought, the mix is exteriorized and thus brings unity to an inaudible intellectual life by providing an atmospheric dispensary for tangents.

This phenomenal modulation would have become more complex for Burroughs if the radios wandered from static toward tuning. Depending upon the density of the dial, information would be taunted heterodynamically by static fusing and warbling along the trajectories of signals, then multiplied by the three radios, to equal an axial formation splayed across the room—a drifting din between and among many stations at once. Because all would be disruption it would provide a silence where there could be no disruption. Is this odd bucolia why artists and composers have historically tuned in between stations, or why inbetween stations has been a youthful entreaty into art? Pauline Oliveros recounted how, growing up in the 1930s,

> I used to listen to my grandfather's crystal radio over earphones. I loved the crackling static. [...] I used to spend a lot of time tuning my father's radio, especially to the whistles and white noise between the stations. [...] I loved all the negative operant phenomena of systems. (1973:246)

Stefan Themerson's experience took place in the previous decade:

> When I was 14 (in 1924) I built myself a wireless-set [...]. [W]hat fascinated me more than the fact of hearing a girl's singing voice coming to my earphones from such strange places as Hilversum, was the *noise*, to me the Noise of the Celestial Spheres, and the divine interference-whistling when tuning. It became an instrument for producing new, hitherto unheard sounds, which at the time no person would have thought had anything to do with "music." (1970:n.p.)

Maurice Martenot, while a wireless operator at the end of World War I, heard the heterodynes he would later design into the Ondes Martenot of 1928; and in 1933, in the Italian Futurist manifesto *La Radia*, F.T. Marinetti and Pino Masnata proposed "[t]he utilization of interference between stations and of the birth and evanescence of the sounds" ([1933] 1992:268). Among the sounds John Cage wanted "to capture and control," to train for the future of music as he saw it in 1937, was the "static between the stations" (1961:3). Whereas Cage used music to make noise significant (his first music radio foray being in 1942 with *Credo in Us*), a *Newsweek* music critic in 1954 used radio noise to make music insignificant: "Christian Wolff's 'Suite by Chance' could have resulted from Dennis the Menace let loose with an amateur short-wave set" (Dumm 1954:76).

Second Receiver, Third Piece of Furniture

Another piece of furniture often found in many of Burroughs's rooms, from Texas to Tangier, was about the size of a small telephone booth. This orgone accumulator was built according to the instructions of Wilhelm Reich, whose biopsychiatric theory extended the electrical functioning of earlier organismic theories to include a class of bionic energy understood in primarily sexual terms: "There seems to exist *one* basic law that governs the total organism, in addition to governing its autonomic organs. [...] *The orgasm formula [...] emerges as the life formula itself*" ([1948] 1973:5). Orgasmic energy was at play between inorganic and organic states, sparking and tingling inside and outside the organism and, most importantly, it was distributed throughout the earth's atmosphere, an eroticized Bachelardian logosphere gone past the talking stage. The orgone box was designed to receive and concentrate this energy and to pass it on to the individual seated inside. The dissipation and accumulation of orgonotic energy between the individual and the atmosphere was thereby the fundamental, global exchange of life energies, a way of situating the seated.

Although Reich had invented the accumulator before 1945, by the time Burroughs began soaking up orgone energy it was set against the background of another radiation: from the bombs the United States exploded on the citizens of Hiroshima and Nagasaki and from the above-ground testing that followed during the postwar years. The atmosphere was now radiant with orgone energy and fallout, not to mention saturated with the transmissions of the consciousness industry in the form of radio and television. In 1950 Joan Burroughs had convinced her husband that atomic fallout was not merely degenerative physiologically but was also involved in psychic control.[1] Five years later the effects of above-ground nuclear tests conducted by "these life-hating, character armadillos" (Burroughs's Reichian slang for severely repressed individuals) were very much on his mind: "Thirty more explosions and we've had it, and nobody shows any indication of curtailing their precious experiments."[2] In 1957 Burroughs read the "most sinister news bulletin" that reported that the "only forms of life that mutate favorably under radiation are the smallest, namely the viruses. Flash. Centipedes a hundred feet long eaten by viruses big as bed bugs under a gray sky of fall-out,"[3] and he thought that a virus in Tangier that purportedly suppressed the sex drive might be one such mutation: "God knows how many atypical virus strains may follow in the wake of atomic experiments."[4] In *Interzone* the imagery of atomic mutations combined with the radiation technology of the orgone accumulator to produce the variety of mutants in the famed "Spare Ass Annie" section: "Pregnant women were placed in the boxes and left on the peak for a period of three hours. Often the women died, but those who survived usually produced monsters" (1989:102–03). People play radio throughout the day as a background sound track to their anomie and to pathetically establish themselves among a serial community. The community is atomized, and radio static, the sound of a Geiger counter.

Soft Rock Crystal Set

But can Burroughs's three radios be reduced to generators of a productive noise, musicalized sound, instrumental registration, or set up as a private sensory clinic? Despite how "off-tuned" they were, they can never be less significant than a potentiated broadcast. In this way, they stood next to the radio stationed at the very beginning of the style made famous in *Naked Lunch*, a style first exercised in the *Interzone* piece known as "Word." In the beginning was "Word" and the word was radio from its beginning, in the form of a descriptive pastiche of sounds that audibly carries words off the page at the moment one arrives on the page. In a similar action, the sounds were heard on a receiver that was ejaculating.

> The Word is divided into units which be all in one piece and should be so taken, but the pieces can be had in any order being tied up back and forth in and out fore and aft like an innaresting sex arrangement. This book spill off the page in all directions, kaleidoscope of vistas, medley of tunes and street noises, farts and riot yipes and the slamming steel shutters of commerce, screams of pain and pathos and screams plain pathic, copulating cats and outraged squawk of the displaced Bull-head, prophetic mutterings of *brujo* in nutmeg trance, snapping necks and screaming mandrakes, sigh of orgasm, heroin silent as the dawn in thirsty cells, Radio Cairo screaming like a berserk tobacco auction, and flutes of Ramadan fanning the sick junky like a gentle lush worker in the gray subway dawn, feeling with delicate fingers for the green folding crackle.
>
> This is Revelation and Prophecy of what I can pick up without FM on my 1920 crystal set with antennae of jissom. (Burroughs 1989:135–36 and 1959:229)

For Burroughs there were two forms of technology enabling an ease of ejaculation. The first is the chemical by-product of junk: the hair-trigger masturbation while kicking a habit. The second is the orgone box: "The orgones produce a prickling sensation frequently associated with erotic stimulation and spontaneous orgasm. —Now a spontaneous, waking orgasm is a rare occurrence even in adolescence. Only one I ever experienced was in the orgone accumulator I made in Texas."[5] But the jissom antennae are part of a much more complex technology. Jissom is made of the same protoplasm which Count Alfred Korzybski, in his organismic theory of general semantics championed by Burroughs, characterized as "human being." Protoplasm not only mobilizes all the organism's psychophysiological functioning, it also connects it colloidally with inorganic matter. Its radiophonic significance comes into play first in the way that the surface is not merely on the surface, but enveloped evenly throughout, and second because "by necessity all surfaces are made up of electrical charges" (Korzybski [1933] 1958:113). The greater the surface, the better the reception. The more accelerated the transmission of stimuli throughout the organism via the protoplasmic medium, the more pronounced the psychosomatic effect upon consciousness.

The first sound Burroughs heard on this technology—*schlupp*—arose from the love Burroughs had for Allen Ginsberg toward the end of 1953. According to Ginsberg, "Schlupp for him was originally a very tender emotional direction, a desire to merge with a love, and as such, pretty vulnerable, tenderhearted and open on Burroughs's part" (in Miles 1990:155). Schlupp has a cartoonlike onomatopoeic relationship to sounds of saliva, sweat, semen, and other sexual fluids and to the sounds of eating and digestion—all bodily sounds which, with the exception of the presence of teeth, have no bones. It is an appropriate sound for the unhewn hungers of junk or sex, for it is the body's interior making its needs conspicuously known within the world. In his writing, schlupping first appeared as raw lust, "an amoeboid protoplasmic projection, straining with a blind worm hunger to enter the other's body, to breathe with his lungs, see with his eyes, learn the feel of his viscera and genitals" (Burroughs 1985b:36). Homosexual desire here produces an image of the cohabitation of one body since, according to Burroughs, "It's a crucial factor in homosexual relationships to be the other person" (in Bockris 1982:60).[6]

But schlupp quickly became destructive and pathogenic, being the sound effect for a full-body disembowelment and for a slimy, junk-driven osmotic action that completely *usurps* the other person. That was not the only sound of junk. A junky's hunger could betray itself through privatized ultrasound or radiophonic transmissions with a "black insect laughter that seemed to serve some obscure function of orientation like a bat's squeak," or that could force the junky to "listen [...] down into himself" to tune in to the "silent frequency of junk" (Burroughs 1959:51). With his beloved substance finally in hand, the junky's "face dissolved. His mouth undulated forward on a long tube and sucked in the black fuzz, vibrating in supersonic peristalsis disappeared in a silent, pink explosion" (52). These are a mix of sound and signals, out of reach of the frequency range of normal communication, feeding on the visual static of black fuzz in order to become stationed.

Burroughs's 1920 crystal set was tuned-in to the "Composite City," an auditive mosaic, a combination of the cultural klatch of Tangier (an arguably neutral zone of complete international intrigue) and the scattered array of correspondence and fragments compressed into service as the manuscript for *Naked Lunch*. The Islamic *din* is heard in "flutes of Ramadan" while the "riot yipes and the slamming steel shutters of commerce" are those of a *jihad*, derived from a little Broadway musical "number called the Jihad Jitters":

> Start is we hear riot noises in the distance. Ever hear it? It's terrific. [...] You wouldn't believe such noises could result from humans, all sorts of strange yips. Then the sound of shop shutters slamming down. Then the vocal comes on.[7]

Schlupp returns as jihad through a mapping of orientalist alterity onto the *alter idem* of Burroughs's dyadic sexual identification: "if someone starts inundating an area with Identical Replicas, everyone knows what is going on. The other citizens are subject to declare a "Schluppit" (wholesale massacre of all identifiable replicas)" (1959:164). The uniform saturation of difference in radio's Composite City is experienced as a complete environment of noise and exemplified by the white noise of radio static, i.e., all possible frequencies at once, the routes through the publicness of the market pathologized and channeled into a rampaging broadcast.

Burroughs experienced this concretely in the noise of the languages foreign to him in Tangier. One day two friends and he were in the city:

> Walking ahead of us was a middle-aged Arab couple, obviously poor country people down from the mountains. And one turned to the other and said, "WHAT ARE YOU GOING TO DO?" We all heard it. Perhaps the Arab words just happened to sound like that. Perhaps it was a case of consensual scanning. (1985a:54–55)

Another friend sought friendship in foreignness through tuning in to interpolation on the radio:

> I had a friend who went "mad" in Tangier. He was scanning out personal messages from Arab broadcasts. This is the more subjective phenomenon of personal scanning patterns. I say "more" rather than pose the either/or subjective/objective alternative, since all phenomena are both subjective *and* objective. He was, after all, listening to radio broadcasts. (1985a:55)

In this way we can ask whether the static on the three radios was nothing but a live feed of the city that housed his room and, in turn, whether the same could be said of all radio broadcasts. For Burroughs, the Composite City local broadcast hallucinated yage-like the entire world as a place "where all human potentials are spread out in a vast silent market" (1959:106).

Two-Way Radio

The Composite City was also the myriad of fragments from correspondence and other sources held together as material for the manuscript to *Naked Lunch*. Many of these letters were trans-Atlantic letters to Ginsberg in which he would write down his "routines," the performative means through which Burroughs generated his ideas. Once, because Ginsberg hadn't written back in a long time, Burroughs became very distressed. His love for Ginsberg could go without the body but he needed someone to listen who understood him completely, so he could reproduce this understanding in himself through autoingestion. This had long been the case and, as Ginsberg said,

> Bill became more and more demanding that there be some kind of mental schlupp. It had gone beyond the point of being humorous and playful. It seemed that Bill was demanding it for real. Bill wanted a relationship where there were no holds barred; to achieve an ultimate telepathic union of souls. (in Miles 1993:66)

Burroughs pleaded with the absent Ginsberg:

> I have to have receiver for routine. If there is no one there to receive it, routine turns back on me like homeless curse and tears me apart, grows more and more insane (literal growth like cancer) and impossible, and fragmentary like berserk pin-ball machine and I am screaming: "Stop it! Stop it!"[8]

He also needed Ginsberg because there was a danger in becoming the type of solitary sender that was later described in *Naked Lunch*.

> A telepathic sender has to send all the time. He can never receive, because if he receives that means someone else has feelings of his own could louse up his continuity. The sender has to send all the time, but he can't ever recharge himself by contact. Sooner or later he's got no feelings to send. You can't have feelings alone. (1959:163)

Sending is also dangerous because it can be debilitated by a lack of response and recoil into various means of control to survive. "Telepathy is not, by its nature, a one-way process. To attempt to set up a one-way telepathic broadcast must be regarded as an unqualified evil" (Burroughs 1959:167). The true evil, however, was not to be expressed in interpersonal ways but was itself an expression of the control exerted by the authoritarian state. In other words, the weight of humiliation of unrequited love has rolled over from a total lack of due process into outright manipulation and, moreover, it finds its technology hot-wired in encephalographic research: "Shortly after birth a surgeon could install connections in the brain. A miniature radio receiver could be plugged in and the subject controlled from State-controlled transmitters" (163). Two-way interpersonal communications become the means of both inner emigration and outright resistance against impersonal authority and the depersonalized crowd which it produced: "telepathy properly used and understood could be the ultimate defense against any form of organized coercion or tyranny on the part of pressure groups or individual control addicts" (167). For Burroughs, just as the pathogenic slide into sending could be broken by one letter from Ginsberg (more specifically, letters at intervals to satisfy his routine habit), so too could authoritarianism be broken by actual communication. In this respect, he echoes Korzybski's distinction between "sanity and un-sanity" subsumed within his overarching category of *Physics and Related Sciences*:

> *Non-aristotelian, Scientific, Adult Standards of Evaluation*:
> Radio, as powerful means of communication and education.
>
> *Aristotelian, Infantile Standards, Evaluation of Commercialism, Militarism*:
> Commercialized radio, advertisements, private propaganda, often stimulating morbid inclinations of the mob. (Korzybski [1933] 1958:557)

Three Oft-Tuned Radios

The three radios in his room in Tangier, the orgone box, and the 1920 crystal set with antennae of jissom were all receivers. For Burroughs, someone sending—anything—meant someone had to be able to receive it correctly or it could transform into a form of control. Without a built-in response his love and routines could make him sick, lost, and alone in the Composite City. To receive correctly, the person had to be predisposed through similarity for communication and not have similarity imposed upon them. In other words, it was a communicative relationship based upon the merger of individuals into a nonpathogenic schlupp, not into a crowd that has internalized a common set of supple dictates. Likewise, receiving could be therapeutic only if it contained the proper radiation; it could situate him among the global exchanges freely occurring within an openly sexualized atmosphere. Yet this was forced to occur against a background radiation of atomization, dissolution, and transmission whose advance into the foreground meant total control and destruction. The static of the three off-tuned radios, in this respect, was both communication and radiation. It was a potentiated broadcast situated in between sending and receiving, a place where Burroughs could listen and hear something similar to himself collected within the white noise of otherness. And no threat could come from an impotence engineered by a refusal to tune in.

Notes

1. From a letter to Allen Ginsberg, 1 May 1950 (in Burroughs 1993:70).
2. From a letter to Allen Ginsberg, 9 January 1955 (in Burroughs 1993:254).
3. From a letter to Allen Ginsberg, 15 June 1957 (in Burroughs 1993:359).
4. From a letter to Jack Kerouac, 4 December 1957 (in Burroughs 1993:379).
5. From a letter to Allen Ginsberg, 16 September 1956 (in Burroughs 1993:326; and in Burroughs 1964:136).
6. During the same conversation, Burroughs elaborated:

   > In homosexual sex you know exactly what the other person is feeling, so you are identifying with the other person completely. In heterosexual sex you have no idea what the other person is feeling. [...Y]ou can identify with them to the extent that you become them, which of course is quite impossible with heterosexual sex because you're not a woman therefore you cannot feel or know what a woman feels. (in Bockris 1982:60)

7. From a letter to Allen Ginsberg, 29 October 1956 (in Burroughs 1993:339).
8. From a letter to Allen Ginsberg, 7 April 1954 (in Burroughs 1993:201).

References

Benjamin, Walter

1979 "One-Way Street." In *One-Way Street and Other Writings*. Translated by Edmund Jephcott and Kingsley Shorter. London: New Left Books.

Bockris, Victor

1982 *A Report from the Bunker with William Burroughs*. London: Vermillion.

Burroughs, William S.

1959 *Naked Lunch*. New York: Grove Press.

1964 *Nova Express*. New York: Grove Press.

1985a "It Belongs to the Cucumbers." In *The Adding Machine*. London: Calder.

1985b *Queer*. New York: Viking Penguin.

1989 *Interzone*. New York: Viking Penguin.

1993 *The Letters of William S. Burroughs: 1945–1959*. Edited by Oliver Harris. New York: Viking Penguin.

Cage, John

1961 "The Future of Music: Credo." In *Silence: Lectures and Writings*, 3–6. Middletown: Wesleyan University Press.

Dumm, Robert

1954 "Sound Stuff." *Newsweek* 43 (11 January):76.

Korzybski, Alfred

1958 [1933] *Science and Sanity*. Lakeville, CT: The Institute of General Semantics.

Leary, Timothy

1983 *Flashbacks: A Personal and Cultural History of an Era; An Autobiography*. Los Angeles: J.P. Tarcher.

Marinetti, F.T., and Pino Masnata

1992 [1933] "La Radia." Translated by Stephen Sartarelli. In *Wireless Imagination: Sound, Radio and the Avant-Garde*, edited by Douglas Kahn and Gregory Whitehead, 265–68. Cambridge, MA: MIT Press.

Miles, Barry

1990 *Ginsberg: A Biography*. London: Penguin Books.

1993 *William Burroughs: El Hombre Invisible; A Portrait*. New York: Hyperion.

Oliveros, Pauline

1973 "Valentine." Appended to *Electronic Music* by Elliott Schwartz. New York: Praeger.

Reich, Wilhelm

1973 [1948] *The Cancer Biopathy*. Translated by Andrew White, et al. New York: Farrar, Straus and Giroux.

Themerson, Stefan

1970 "Letter to Henri Chopin." *Ou* 36/37:n.p.

INAUDIBLE POST*SCRIPT*

A silent coda on the **disembodied** voice and the subsequent **un**writing of history

Rev. Dwight Frizzell and Jay Mandeville

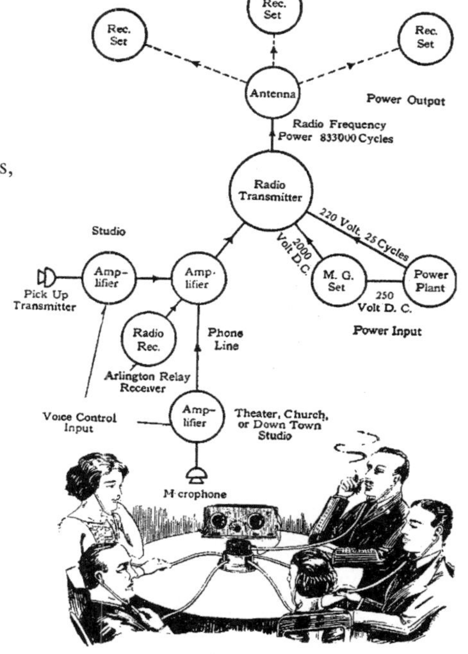

Easy to erect radio antenna towers,

Jefferson tube rejuvenators,

the WARREN LOOP,

Bakelite Philcos,

Super-heterodyne Radiolas,

Dutho battery eliminators,

self-contained **Radaks**,

Miraco Ultra 5s,

Mu-Rad portables,

Amplion Dragon horn speakers,
and the early Deco Operadio were the eagerly sought after and often purchased hardware in the emerging snakelike Medusoid radiophonic marketplace.

What the populace *really* wanted:

- ❏ improved reception of the ubiquitous potted-palm music,
- ❏ vaudevillian smear tactics,
- ❏ sopranified "Blow Blow Winter Wind" pseudo-classical recitations,

❑ and the low-slung rinkydink **xylophone muzakifications** that dominated radio's earlier eras.

Even then,
one **twist** of the dial and your Philco Bakelite bone oracle would advise you, with maddening **acausal** parallelism, to simultaneously soap up,

brush off,

drip down,

ship out,

PUMP UP,

SEND FLOWERS,

eat wheat,

smoke cigars,

BUY LICORICE,

invest in war bonds

and love thy neighbor,

as radiomongers attempted to shatter
the Jerichovian mind-over-chatter barrier
with a **trumpeting** barrage
of singing advertisements,
woozy Winchellisms,
and **jazz-coded** rag-to-riches morality plays.

The culture-wide **myth realignment**
that resulted from this **ceaseless**
consumerization of the listener
came about semi-somnambulistically,
reprioritizing listener preferences toward
the sudden, the ultra-dramatic,

the over-per+cus+sive and the superloud.

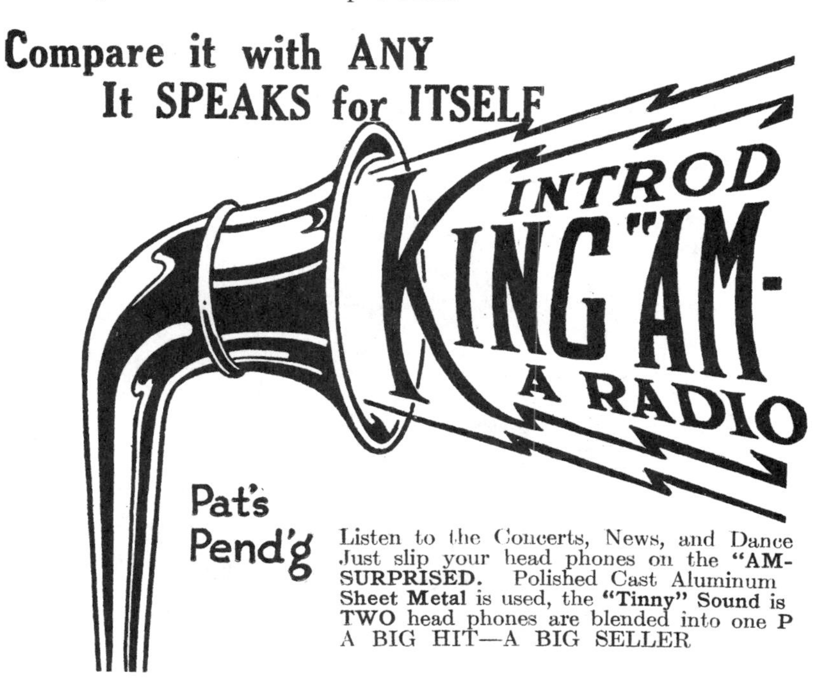

The so-**called** "sponsors' message"
l a g g e d far behind the snappy presentation techniques.

Radio fans aligned themselves
according to their own whimsical metaprogrammatic inclinations,
responding to radio in ways that **short-circuited**
advertisers' expectations,
imposing individualistic desires
and intimate iconographies
in an overlay of psycho-structures
that employed the moment-to-moment radio soundscapes
as a mere **backdrop** to highly idiosyncratic architectonics,
creating a personal theatre
in which the actual production takes place in the living room,
bedroom, or wherever the receiver is made to articulate

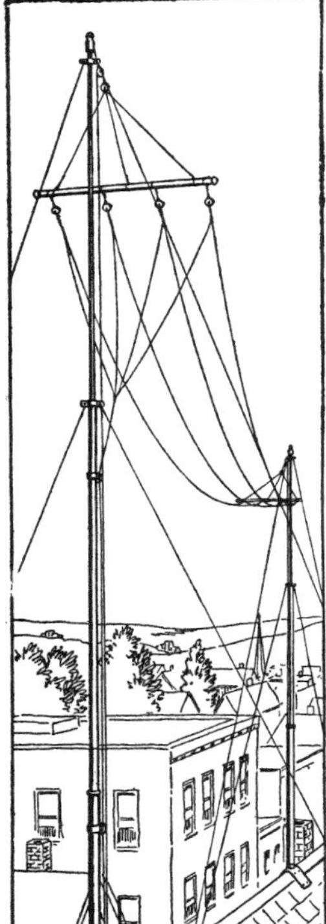

those ever-wa n d e r i n g waves.

The worm-holing of spatial constraints
that allows us to scope out (with techno-omniscience)
the tribulations of the Pope in Moscow,
student voices from Tiananmen Square,
the roar of the astro-pundits drifting
near the apex of Houston's famous dome,
and the redoubtable **rumbling**
of modern French poets
amid the expanding and contracting
soundboards of Paris's **IRCAM** studios
has miraculously and without precedence
placed us at sea amid a **crackling** torrent
of audible presences that project themselves
across the narrow electro-magnetic bandwidth
we cock our ears to.

Materiality, sensuality, fleshed-out lips, guttural mutters, the seemingly
anonymous caress of thousands of voices
poised **inch**e**s** from the ear muzzle us
with their musings, immerse us
in the grainy configurations of body-talk

versus the merciless **disembodied** tongue.

These are the phenomena that trigger
a rebounding internal plenitude of unschematizable
atemporal events that enable us to scrutinize
our own tiny corner of the ecosphere more closely
and successfully elicit its *secret utterances*
amid the air-borne hard-sell strategies
behind broadcast **bombast**.

As always, with radio,
what you **hear** is what you see—
the relationship between the observed and the observer
is radically altered,
the *virgin ear* is immersed in <u>democratic</u> <u>clarity</u>.
By reducing the necessity for the cramped,
spidery trail of written record-keeping,
radio is, essentially, **un**writing its own history.

The ghost in radio's machinations
is perpetually bifurcating
and attempting to square itself.
The modern gustatory devouring of voice
or music by tape or digital optics
and its subsequent **regurgitation**
over the airwaves
allows us to continually reinitiate the flow of the audible body
and exteriorize consciousness.

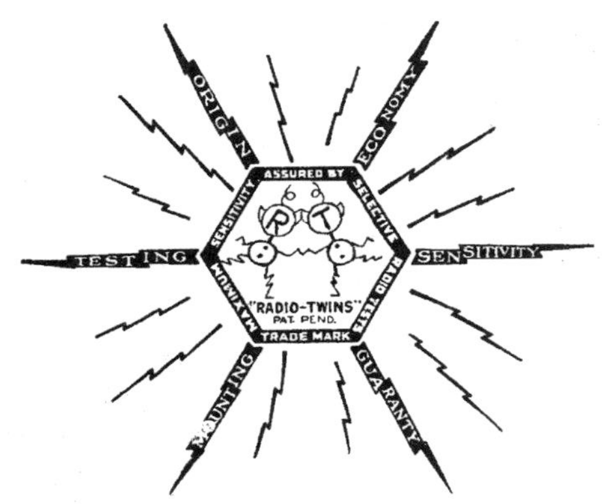

Even before the wireless message reaches us,
the obscure prompting of our **egoless** selfhood
helps us escape the potentially laborious reality
interpretations of over-the-air talents
and promotes the **ecstatic** rise
of multiple **selves**.

"Radio - Twins"
"HEXAGON"
SUPER-CRYSTALS
Have Six Points of Superiority

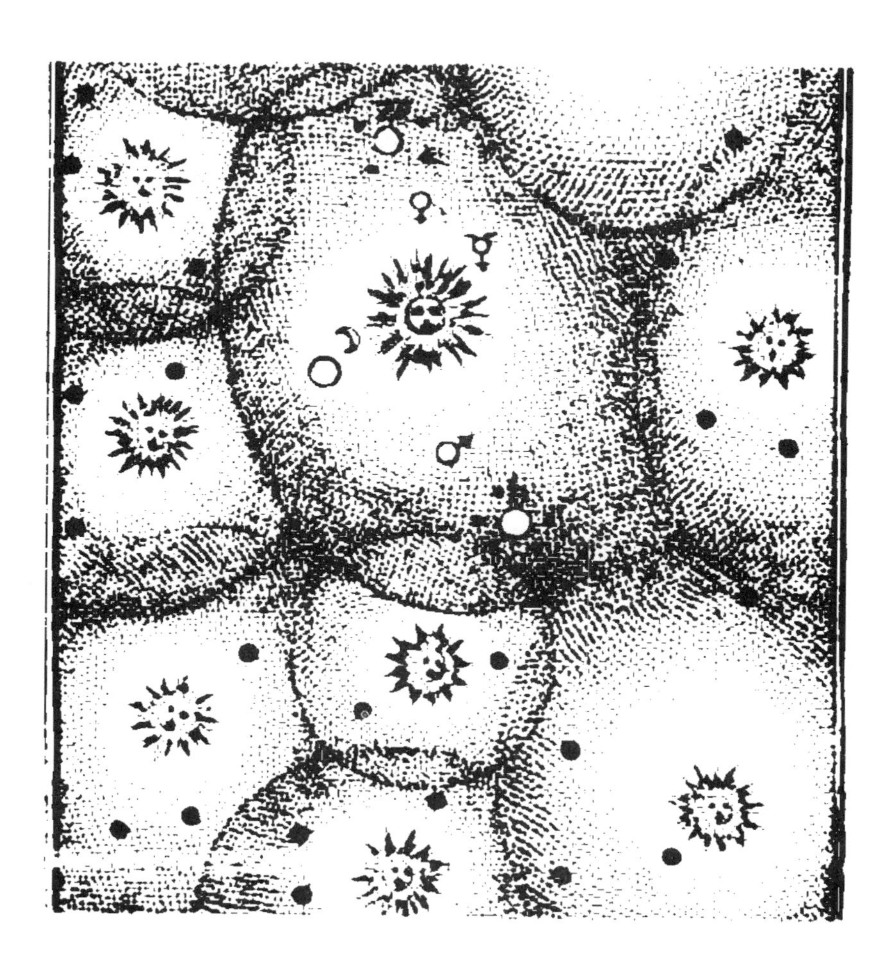

The clash of differentiation with uniformity persists
in every radio listener until we are delivered
from the Kierkegaardian panic of solitude
by **uplinking** with the emerging potentia state
that could be called the **Broadcast Omniverse.**

The full colonization
of the Broadcast Omniverse
by the widest spectrum of imaginative humanity
is the *raison d'être* of radio band phenomenology,
an audible staging that allows human action
to **synchronize** with its own cephalic thrusts
and noetic assertions.

The familiar radio apparatus **tunes us** in
to our own emotional manifestations
by putting us on that more omniscient wavelength,
capturing us in etheric amber
as we pause before entering
the limitless time frame of **incessant** Babel.

Radio Play Is No Place

A Conversation between Jérôme Noetinger and Gregory Whitehead

Gregory Whitehead

NOETINGER: From your perspective, radio is more than just a vehicle for transmission of sound art: it offers its own autonomous space, its own material?

WHITEHEAD: Absolutely. I strongly believe that radiomakers must find ways to disrupt the boundaries of "sound art," most of which sounds very tired and familiar anyway. Radio happens in sound, but I don't believe that sound is what matters about radio, or any of the acoustic media. What does matter is the play among relationships: between bodies and antibodies, hosts and parasites, pure noise and irresistible fact, all in a strange parade, destination unknown, fragile, uncertain. Once you make the shift from the material of sound to the material of the media, the possibilities open to infinity, and things start getting interesting again. Each broadcast takes place inside an echo chamber of informations, histories, biographies, life stories—and inside the echo chamber resounds the most unnerving question of all, the ghost question: Who's there? Is anybody out there on the other side of the wall, on the other side of this broadcast? Of all the questions that have rattled around inside my head over the past 10 years, that is the most persistent. So radio is certainly most captivating as a place, but a place of constantly shifting borders and multiple identities, a no place where the living can dance with the dead, where voices can gather, mix, become something else, and then disappear into the night—degenerates in dreamland.

NOETINGER: Does a radio work of this sort exist if not heard by an audience? What about the pure play of the radio waves themselves?

WHITEHEAD: Yes, this is the uncertainty that hangs over any broadcast. You cannot know in advance which kind of "play" you are going to transmit. Until, that is, you get something back: a phone call, a postcard, a shout in the dark. A censor. Or a silence. Is the circle completed, or does it gape open, only a theory? As for the "pure" play of the waves, radiowaves by themselves—I suppose one could make something interesting from such purity, but to my ears radio waves fascinate because they are so dirty, that is, the airwaves are so full of voices and bodies trying, in one form or another, to get into the ears of somebody else. Stripped of its raucous Babel of attempted and aborted contacts, radio becomes just another noisemaker, and we already have plenty of those. That's why I have never been impressed by various art-radio projects that simply play with or recontextualize existing signals: unless you are willing to electrify yourself and enter directly into the flow of relations, the Limbo Zone of transmission, then you're not really doing anything more than pushing buttons, and that just isn't enough anymore. On the other hand, the play *between* signals carries its own fascinations: When I was 10 or 11 years old, I would lay in bed with a shortwave radio under my pillow, slowly turning the dial, searching for the weird signals between the stations, composite voices, strange languages collapsing into each other. Years later, I learned of the theories of Konstantin Raudive, who believed that these between-zones were assembly halls for the voices of the dead.

NOETINGER: Many of your radio works have also been released on CD—isn't there a contradiction here? For me, radio means only one listening, and with a CD you can listen as much as you want. Maybe you are more of an owner with the CD than with the radio; and when you know you can listen only once, maybe you pay more attention.

WHITEHEAD: Do we really want to fix media identities so strictly? To my mind, what is interesting is the way media circuits cross, evading format borders, or putting them into question. In the Theatre of Operations, this becomes explicit in the attempt to incorporate the idea of "circuit" into the performance of the piece itself, implicating all kinds of materials and contexts—stick a needle in the brain, and spin those tunes. One story: a few years ago, a convict in San Quentin Prison contacted me for a cassette copy of *The Pleasure of Ruins* [1988]. He had heard it played on KPLA, in San Francisco, and could not believe his ears. So I immediately sent him a copy, and he sent a letter of thanks back, telling me that he and a few buddies were using it as an exercise tape. OK, I thought, hey, there's a direct, practical use I had not anticipated. Then about a year ago, by sheer coincidence, I met the lawyer who was representing him before the California Board of Appeals. I told her the story, and she laughed, asking me how well I knew prison argot. Not well enough, it seems, because then she told me that in San Quentinese, "exercise" means "masturbate." So here is an example of a complex circuit of communications running from radio to prison to telephone (calling the station) to post office to cassette to individual nervous systems. Such improbable and unpredictable circulations among institutions, media, and bodies are part of what gives life to a work, the transmission taking on a kind of itinerary.

NOETINGER: Why the frequent references in your work to Artaud?

WHITEHEAD: One of my first experiences with radio performance was a doomed attempt to give a simultaneous translation of Artaud's *Pour en finir avec le jugement de Dieu* [recorded in 1947]. Not just the text, but also the intense unearthly quality of voice, through all of its entranced and wild gyrations. Like the voice of radio, Artaud's voice is literally all over the place: talk-show, tirade, incantation, threat, confession, lament. Beyond that personal experience of Artaud inhabitation, I have always found the piece emblematic of a very compelling, stripped-down form of radio, a form of "poor" radio (in the Grotowskian sense of "poor theatre"), the direct confrontation of a body poli-

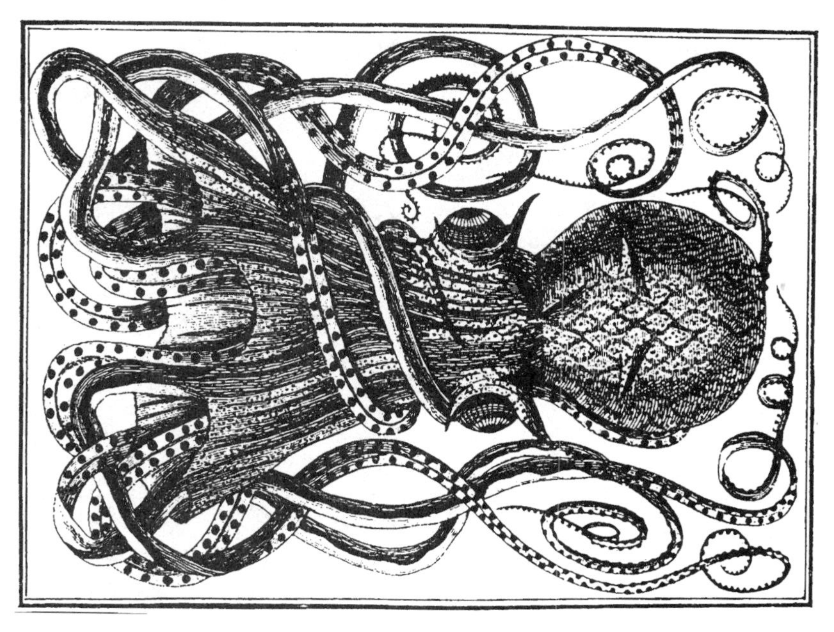

1. *A proposal for the future of electronic media.*

tic with the contusions and contortions of a body alone, one nervous system to another, a form that remains tremendously appealing to me. Of course, the prospect of such an electrified confrontation made the director of French radio (a man named Vlad Porché), so nervous that he canceled the broadcast, and it was not heard in France until 1973. I also hear *Pour en finir...* in relation to another emblematic work, Orson Welles's *War of the Worlds* [1938]. From a war raging inside one man's brain, we switch to an alien invasion. Yet the experience of shock, and the sensation of airwaves suddenly "taken over" by The Other (Artaud: Le Mômo, who hails from the Bardo Zone; Welles: alien invaders from Mars) remains constant. Unlike *Pour en finir*, the Welles *War* became possibly the most notorious broadcast in history, creating panic in the streets. Nonetheless, it was also a kind of "poor" radio, a simple organizing concept surrounded by a few cheap sound effects and a small ensemble of improvising actors.

NOETINGER: You also have done live performances that then find a way into your broadcasts; another kind of circle?

WHITEHEAD: Here again, the key question: who's there? Since I give occasional presentations on issues of technology, language, bodies, brains, publics, programs, and so on, I had the idea of conducting playful exercises with the audience: learning how to speak backwards, the correct way to pronounce "prosthesis," how to speak like an analog degenerate, various conceptual singalongs, and so on. Recordings of these group exercises then become part of the archive for Theatre of Operations plays: one public folded back into another. In an intermedia concept like *Pressures of the Unspeakable*, the audience performs a different role, becoming "scream donors" to an answering machine "scream bank" located at the host station. These screams are allowed to accumulate over several weeks, then are assembled and intercut into a local screamscape, which is then broadcast, with phone lines remaining open. The grand acoustic icon of modernism (the scream) is set loose inside the pinball machine of the postmodern media. The eventual broadcast becomes a catalyst for more scream flow, that is, more calls. The circuit of broadcast and public response could continue, theoretically, indefinitely, though I'm still waiting for a station to permit me to test this. A search for the Last Scream. Once more, the idea of confounding and encircling public with private, immediate with distant, noise with silence, voice with technology, circling back again and again, piling up meanings as fast as old generations fade out, a spiral of communications transforming itself into an improvised community that is always in danger of spinning out of control, losing itself, the idea of The Producer also getting hopelessly and gratefully lost in the vortex, deep in the media screamland blues.

NOETINGER: You mention the Last Scream—but what about the Primal Scream?

WHITEHEAD: Ah yes, the Scream of Screams. I'm not sure I'd know it if I heard it, or even if I screamed it. Possibly the most primal scream we can ever know, hanging at the end of the millennium, is the electrified white noise cry of whole communities suspended on the brink of extinction. A primal scream that is also a death rattle. Or maybe, American Talk Radio, a different category of white noise.

NOETINGER: You often talk about "relationships" in your work: what about the kind of relationship McLuhan talks about with regards to the Global Village?

WHITEHEAD: Right, the glorious, glowing Global Village, which to my mind is sort of in the same elusive category as the Primal Scream. The problem with the whole constellation of ideas having to do with the electronic tribe—radio as talking drum, the wired society, the Neural Net—is that there is no necessary or automatic relationship between communications technologies and community. The slogan that "communication equals community" is only true when people are willing to work very hard to achieve it, and are then willing to fight to preserve the fragile community they have built. There is a utopian aspiration in all communication technologies, but the utopian side is counter-balanced and all too often canceled out by the darker drive, the connection between information and war, between communication and the command or control over communities. This is the other side of Radio Utopia: *Radio Thanatos*, and I hear it more now than ever, whether in Sarajevo, China, or in the streets of Los Angeles. The root for "utopia" is the Greek *ou topos*, or "no place." And radio is perhaps the most powerful and destructive No Place ever conceived or conjured.

NOETINGER: You often use your own voice as the principle and sometimes only sound source: what is this "voice"?

WHITEHEAD: A question with several answers: To begin with, I've always been uneasy, or maybe just plain bored, with the phonocentric tradition of sound poetry, in which the voice becomes an onanistic fetish-object with which to explore the subjectivity of the one who speaks. To my ears, work in this tradition typically flattens out everything that is distinctive in an individual voice, all the things that do not add up to The Real Person, because in the saturated buzz-world of electronic media, our voices are inscribed with all kinds of "phonies" other than our own. The fact is, we cannot find our voice just by using it: we must be willing to cut it out of our throats, put it on the autopsy table, isolate and savor the various quirks and pathologies, then stitch it back together and see what happens. The voice, then, not as something which is found, but as something which is written. We may have escaped from the judgment of God, but we have not yet escaped from the judgment of the Autopsist—the truth is not in how your voice sounds, but in how it's cut. If we want to find our "real" voice, we must be prepared to figure and refigure. Such is the Postmortem

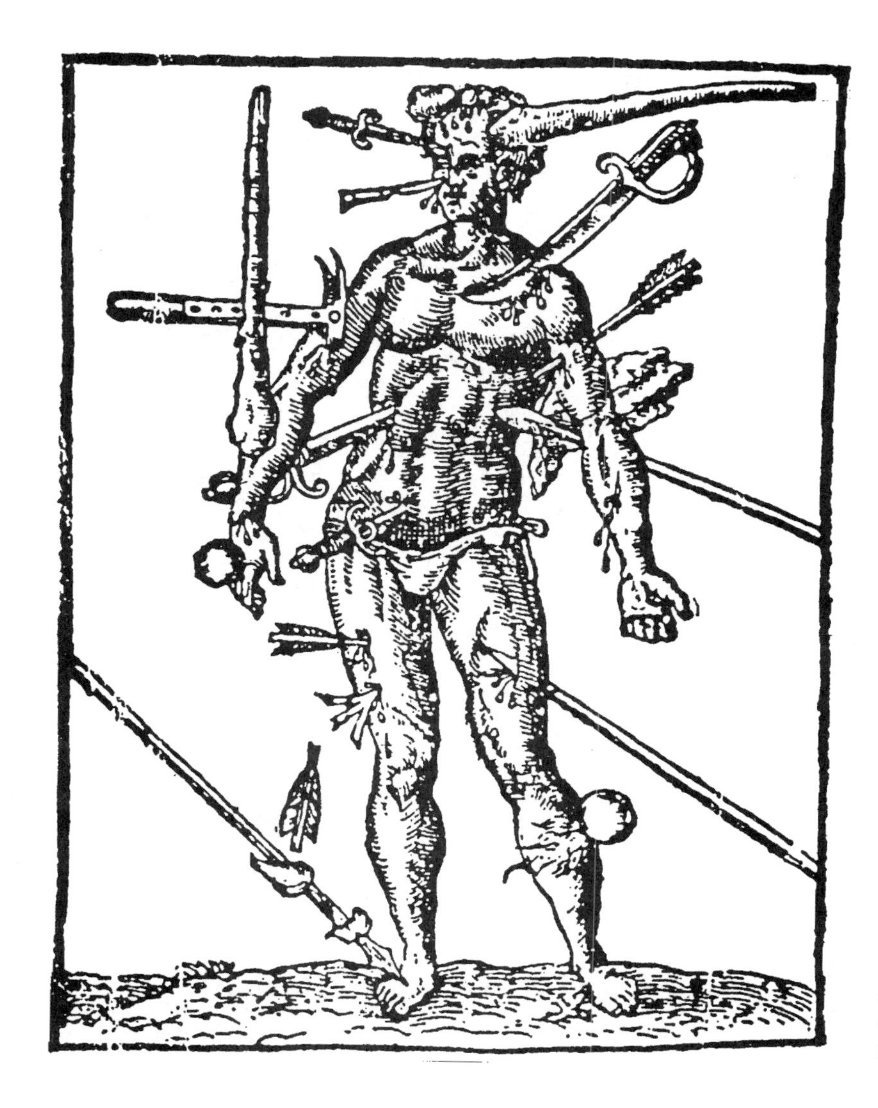

2. *Wounds can bleed or they can sing: the difference is a matter of technique.*

Condition. Further, the problem of voice raises, inevitably, the problem of bodies, so working with voices of every category and derivation provokes questions of politics at the most microscopic and essential level, a politics of positioning another's body. Fortunately, I soon discovered that the problem of voicebodies (and the hunger to become entangled with other voicebodies) could resolve itself into the pure pleasure of speech in ruins. That is, the prosthesis can be a twitching finger of ecstasy as much as the trigger finger of death. Wounds can bleed or they can sing: the difference is a matter of technique. And finally, voice in the broadest sense, or the position of the auteur. Here, I'm very attracted to the idea of establishing a concept, and perhaps a set of procedures, and then removing myself from the loop, letting the concept take on a life of its own. Along these lines, I've always liked the French word *animateur* with regards to the media, that is, the one who might breathe life into an apparatus, even as an artificial respirator, but who then withdraws. If you need to be in control from start to finish, then in a sense, nothing is happening.

NOETINGER: But also, the hearing of other voices in one's own head, as in schizophrenia, appears to occupy an important place, no?

WHITEHEAD: Well, schizophrenia has its acoustic double in schizophonia, the "split voice." Years ago, I performed a conceptual talk show in which I presented myself as the Director of the Broca Memorial Institute for Schizophonic Behavior. We invited listeners to call in and share with us their schizo/voices, the voices that they heard clamoring about in their heads. Amazing calls flooded in, people speaking in every kind of twisted tongue. Then one woman phoned to tell us, with evident relief, that she had for quite some time thought she was schizophrenic—but after listening to us, she realized she was "only" schizophonic! I have long been fascinated by the case of Louis Wolfson, a man at extreme odds with American English, his mother's tongue. His war against the acoustic oppression of the sound of this language is recorded in his extraordinary *Le Schizo et les langues.* As a defense against such acoustic tyranny, Wolfson became a magician of dissection and reassembly, stitching together a parallel language from the bones and organs of French, Russian, Yiddish, and German—a language of his own that would give him sanctuary from the crushing tonalities of hated American. To achieve this remarkable bit of psycholinguistic montage, Wolfson relied on two main resources. First, dictionaries, offering static and neutral raw material. And second, a resource that perhaps also offered him an alternative mother, the no place that may be the mother of us all: a radio.

Stein's Stein

a tale from *The Aphoristic Theatre*

Allen S. Weiss

Gertrude Stein was already one of William James's favorite students by the time she enrolled in Hugo Münsterberg's laboratory experimentation course during her sophomore year at Radcliffe in 1894. James, having just published his monumental *The Principles of Psychology*, was at that moment particularly interested in the relations between conscious and unconscious states of mind. Among his oblique and eccentric entries into the domain was a study of the spiritualist practice of trance-induced automatic writing, much to the dismay of his more rationalist colleagues. In this light, he initiated a series of psychological experiments in which the subject was tested according to varying degrees of fatigue and distraction. One of his subjects was Miss Stein, whom he was to describe as "the ideal student." She recounts her own experience: *Strange fancies begin to crowd upon her, she feels that the silent pen is writing on and on forever. Her record is there, she cannot escape it and the group about her begin to assume the shape of mocking friends gloating over her imprisoned misery.* Bizarre as this might be, it was in any case better than those experiments conducted by one of her classmates, notably less than ideal, who complained to Professor James that the consistent lack of response by one of the students was falsifying the results of his work. When asked who the student was, he replied that it was Gertrude Stein. The Harvard professor, true to form, found her lack of response to be perfectly normal. Stein was greatly amused by this incident, remembering that after the French Revolution, Saint-Just proposed that in the *lycées* the prize for eloquence be given for laconism. Of course, the tight-lipped New Englanders who were her classmates would certainly understand such a procedure; for her, coming from San Francisco, this was a quite novel approach to things. In any case, in his wisdom, James was much later to explain to his prize student: "Never reject anything. Nothing has been proved."

Inspired by her mentor, Gertrude Stein was immediately to begin a series of supplementary experiments with her friend, Leon Solomons, the results of which were published in the Harvard *Psychological Review* of 1896, entitled "Normal Motor Automatism." This was Gertrude Stein's first publication. There were only two subjects, herself and Solomons. Her description: *This is a very pretty experiment because it is quite easy and the results are very satisfactory. The subject reads in a low voice, and preferably something comparatively uninteresting, while the operator reads to him an interesting story. If he does not go insane during the first few trials he will quickly learn to concentrate his attention fully on what is being read to him, yet go on reading just the same. The reading becomes completely unconscious for periods of as much as a page.* Yet the "he" was a "she," since it was Gertrude who took dictation from her friend, attempting to couple automatic reading and automatic writing. *For this purpose the person writing read aloud while the person dictating listened to the reading. In this way it not infrequently happened that, at intersecting parts of the story, we would have the curious phenomenon of one person unconsciously dictating sentences which the other unconsciously wrote down; both persons meanwhile being absorbed in some thrilling story. Typical of the resulting sentences is: Hence there is no possible way of avoiding what I have spoken of, and if this is not believed by the people of whom you have spoken, then it is not possible to prevent the people of whom you have spoken so glibly...* Enough! This was markedly less than thrilling! Not unexpectedly, Solomons noted her "marked tendency to repetition," yet what else could have been expected, given the nature of the experiment? In any case, Stein explained that *the voice seemed as though that of another person.* She was certainly not repeating *herself.*

Stein soon continued the experiments on her own, and published the results in the *Psychological Review* of 1898, in an article entitled, "Cultivated Motor Automatism: A Study of Character and Its Relation to Attention." This was an attempt to induce automatic writing in a larger group of students, equally divided into male and female. Given her newfound freedom, as well as her consequent growing attention to her female classmates, she felt that it was only just to make the same gender distinctions in her objective studies. Again, Stein was one of her own subjects. The results were similar.

Reading is a somewhat more unconscious activity than writing. Automatic writing may thus occur under dictation, in whatever form the dictation takes—whether as the discourse of an embodied other, of a disembodied spirit, or of the unconscious. Münsterberg's psychotechnology and James's empirical psychology in fact did little to raise psychic automatism to the level of art, adding few stylistic innovations or novel content to older forms of mediumistic expression. The genius of psychotechnology was, paradoxically, to have transformed dictation into automatism. Normally, man dictated, while woman wrote; but Miss Stein was only marginally normal, being ideal. Her genius was to have absorbed both dictation and automatism, and yet become a writer—even to the point of eventually being able to write the autobiography of another. Her genius was equally to have transformed the normal into the cultivated. Gertrude Stein, in these experiments at Harvard, was, in her own words, to have become, *the perfect blanc while someone practices on her as an automaton.* That someone was herself.

Aural Sex

The Female Orgasm in Popular Sound

John Corbett and Terri Kapsalis

In *The Pleasure of the Text* (1975), Roland Barthes defines representation and bliss as mutually incompatible terms. Bliss is the limit of selfhood and the threshold of the text; it runs parallel to and is incommensurable with pleasure. One cannot, according to Barthes's schema, represent bliss since bliss is the destruction of representation. With the experience of rapture or *jouissance*, the codes of orderly rhetorical representation are scrambled and the comfort and safety of interpretation are violently punctured. For Barthes, the site of this disturbance is never mass culture, where any potentially ecstatic repetition is "humiliated repetition" and the shock of bliss is engulfed in a deluge of superficially new fashions. The erotic text appears only in excessive scenarios: "if it is extravagantly repeated, or on the contrary, if it is unexpected, succulent in its newness" (Barthes 1975:42). Bliss interrupts language. An orgasm: the blissed-out sound of broken-down speech.

If we abandon Barthes's anti-mass stance, what do we make of the proliferation of sounds of ecstasy that have been a staple of the pop music world since the 1960s? Specifically, how can we account for the meaning of the many works that include or, more often, center on the female voice simulating sexual bliss? Indeed, with the advent of digital technology and the widespread use of digital sampling in popular music, female sex vocalizations (moans, shrieks, gasps, sighs) have become a staple of dance music from hip hop to Belgian new beat.[1] And outside of the arena of music, in contemporary popular pornographic technologies (phone sex, CD-ROM, virtual reality), soundtracks are currently being produced that utilize these vocalizations in a variety of both nuanced and clichéd ways. Is it possible that underlying the simple discomfort and embarrassment that naturally accompanies the public airing of such graphic sex sounds is a more profound disturbance: a gentle threat to the stability of sensical representation? What happens to this seemingly untenable presentation of bliss when it takes the form of a recording? Is this the pleasurable clawback of ecstasy, containment of rapture, and prevention of total textual loss? Or are pop music sex sounds something harder to pin down, something we experience as unsettling to deep cultural architecture?

A number of hard questions arise from this ubiquitous practice, questions that have long been addressed in terms of film and visual pornography, but questions that take on an aural specificity particular to practices that hinge on recorded sound. Linda Williams has discussed the "money shot" in mainstream porn as "evidence" of male sexual satisfaction, and she has explicated the difficulty encountered when attempting to visually render female orgasm.

> [S]ince "normally" the woman's pleasure is not seen and measured in this same quantitative way as the man's, and since visual pornography also wants to show visual evidence of pleasure, the genre has given rise to the enduring fetish of the male money shot. (1993:185)

It isn't that female pleasure is completely unaccounted for, however; indeed, it has a highly codified status in music and film. As the female counterpart to the visually present ecstatic male, evidence of female sexual pleasure is usually deferred to the aural sphere. Hence, within mainstream pornography and mass culture alike, where male sexual pleasure is accompanied by what Williams calls the "frenzy of the visible," female sexual pleasure is better thought of in terms of a "frenzy of the audible." Sound becomes the proof of female pleasure in the absence of its clear visual demonstration. The quantitative evaluation of male sexual pleasure by means of the money shot ("payoff" measured in amount of ejaculate, force, and distance of stream) may, for female sexual pleasure, be represented in the quality and volume of the female vocalizations. Annie Sprinkle, in her video *Sluts and Goddesses* (1992), plays on this very code by charting an extended series of orgasms, superimposing a graph over a video image of her achieving climaxes. The graph's x-axis measures time, its y-axis measures "orgasmic energy." Not coincidentally, the chart reads like a seismic register of the volume of Sprinkle's vocalizations; her "orgasmic energy" peaks at moments when she screams loudest, while the graph's valleys represent quieter, less vocal interludes. Female sexual energy or "letting go," in this case, is explicitly linked to the "release" of sound, the vocal expression of an inner state.

The recognition of this separate standard of measurement for male and female sexual pleasure (liquid volume vs. sonic volume) is at the center of an ongoing debate, popular and more recently academic, over the status and/or possibility of female ejaculation. In an attempt to draw this conundrum further into the heart of feminist and postfeminist theory, Shannon Bell wrote:

> The ejaculating female body has not acquired much of a feminist voice nor has it been appropriated by feminist discourse. What is the reason for this lacuna in feminist scholarship and for the silencing of the ejaculating female subject? (1991:162)

The word "ejaculate," of course, has a convenient double meaning—vocal ejaculation and sexual ejaculation—that allows Bell to conflate "silencing" (an aural phenomenon) and "erasure" (a visual phenomenon). Though the ejaculating female body has been largely excluded from *visual* representation in pornography, the vocal ejaculations of climaxing women are a prominent, perhaps *the* prominent, feature of representations of female sexual pleasure in mainstream porn and popular culture at large. In a discursive formation that measures female pleasure and performance primarily by how much sound is made, the notion of female ejaculation, as Chris Straayer points out, maps a male-exclusive visuality onto women's bodies (1993:168–72), giving them the chance to have "money shots" of their own. This confounds the traditional marker of sexual difference—ejaculation as the sole domain of men—inciting a flurry of scientific, pseudo-scientific, cultural, and social questionings of the verity, the desirability, and the very physiological possibility of such a thing. Striking a nerve at a very deep level, the question of female ejaculation subtly reveals underlying constructs of the "truth" of male and female orgasm (you can *see* men orgasm; you can *hear* women orgasm), the same kind of truth claims around which the use of sex sounds in film and moreover in music function.

Without a visual image to "anchor" it, the recorded sound of sex and sexual pleasure—for example in popular music or phone sex recordings—

raises a number of fundamental questions about the construction of aural codes for sexuality. In the absence of a synchronous or illustrative visual image, what do recorded female sex vocalizations become *evidence* of? Whose pleasure is being represented? On one hand, these vocalizations are conventionally designed to provide sexual arousal for a male listener. At the same time, like the money shot, such pleasure is derived from the assumption or fantasy that a surrogate partner—with whom the listener may identify—is engaged in sexual activity with the vocalizing woman. This complicated structure of viewer- or listener-identification involves a frequently absent, usually male character. Whereas in film one has visual evidence of the sex act and its culmination, sound recording constantly begs the question of evidence. Is she "really" enjoying herself? Are they "really" having sex? As evidence of the truth of her orgasm and the truth of his/her ability to bring her to orgasm, the listener is offered the sound of uncontrollable female passion. Sound is used to verify her pleasure and his/her prowess.

At the base of an economy of pleasure is a biological truth-claim about the "nature" of women's and men's sexual behavior. Men's pleasure is absolute, irrefutable, and often quiet, while women's pleasure is elusive, questionable, and noisy. This gendered opposition augments another biological construction that configures the male and female orgasmic economies differently: male orgasm is seen as singular and terminating; female orgasm is heard as multiple and renewable. The importance of this singular/multiple dichotomy in the world of sound recordings will become clear later in this essay.

The enforcement of this dichotomy between the spectacularization of male pleasure and the aurality of female pleasure calls into play a problematic with legal implications: In a scopophilic society in which one looks for "eyewitness" accounts (as opposed to mere "hearsay"), what defines aural pornography? What is the legal status of non-language-based sexual sound? The pornographic is defined as that which is seen in images or written in language; in both senses, graphic = written. Thus, federal agencies and consumer advocates can easily police visual obscenity in video images and content obscenity in song lyrics, but they have a much more difficult time defining and prohibiting the use of sex sounds in popular music. Female sex sounds are thus a more viable, less prohibited, and therefore more publicly available form of representation than, for instance, the less ambiguous, more easily recognized money shot. Following this logic, this could be seen as another way of sanctioning and popularizing the construction and circulation of women as the objects of sex, as being "on the market."

At a basic level, then, recorded sex sounds are engaged in, on one hand, the production of an erotics and, on the other, a strict maintenance of gender binarisms. At the same time, sex sounds always work in one way or another in relation to the visual, either by playing on the *absence* of image (allowing sex in places you aren't allowed to see it) or by referencing the visual directly, inciting spectacular fantasy. Aural representations of sexual pleasure therefore enjoy a double standard that allows them to occur in places, including public spaces, that would otherwise ban visual pornography, either cinematic, videographic, or print. For instance, one can hear female orgasm sounds in background music while browsing at a popular clothing store, though the same store would never dare screen porn video loops on in-store monitors.

First Station Break

Instructions for a sex-sounds broadcast:

1. Record sounds on recording device (ooh, ooh, ahh, ahh, ahh, oh yeah, squeaky bedsprings, etc.)

2. Rewind tape, play back, turn up volume
3. Face playback device out various windows as sound plays
4. Move speaker throughout space, placing on floor, against walls
5. Watch for response

Love To Love You, Baby

As early as the 1920s and '30s, several genres of singers turned to the "low moan" for erotic effect. White entertainers like Sophie Tucker ("last of the red hot mamas") and Mae West cooed seductively for male audiences, as can be heard on West's 1933 record "A Guy What Takes His Time." Black blues and vaudeville jazz singers used similar techniques, often incorporating sex sounds into the narrative of the lyrics. Luella Miller's 1927 song "Rattle Snake Groan" and Victoria Spivey's 1934 take of "Moanin' the Blues" both use the same combination of sung moans and snake-penis imagery, as Spivey sings: "Now you talk about that black snake juice/ well you haven't heard no moanin' yet/ Aaaaaall...day long/ And when you hear this moanin' it's moanin' you will never forget."[2]

Female sex sounds came to the hit parade of Western pop music with Serge Gainsbourg and Jane Birkin's major 1969 French success, "Je T'Aime." Since that mythological initiation—which was, at the time of its release, banned from radio in many countries[3]—the pop music world has produced a virtual orgy of like-minded songs, songs that are aimed at a cross-section of mainstream, heterosexual record-buying audiences. Though the industry may target these audiences, this does not account for the actual uses made by nondominant audiences, such as various gay subcultures, who might cross-read such music. Nor does it account for the fact that the music industry might have the savvy and "inside knowledge" to market to those subcultures at the same time as it does dominant audiences. These multiple possibilities for consumption make the market for such music larger and even more diverse, as the presence of Donna Summer's 1975 simultaneous gay and mainstream popular hit "Love to Love You, Baby" attests.

A short list of songs that contain female orgasm sounds includes Marvin Gaye's "You Sure Love to Ball" (1973), the Time's "If the Kid Can't Make You Come" (1984), Duran Duran's "Hungry Like a Wolf" (1982), Prince's "Orgasm" (1995) and "Lady Cab Driver" (1982), Chakachas's "Jungle Fever" (1973), Major Harris's "Love Won't Let Me Wait" (1975), The League of Gentlemen's "HG Wells" (1981), Little Annie's "Give It to Me" (1992), Lee "Scratch" Perry's "Sexy Boss" (1990), Aphrodite's Child's "666" (1972), P.J. Harvey's "The Dancer" (1995), and Lil Louis's "French Kiss" (1989).[4] Within the diegesis of most of these examples, a male lead singer satisfies, either directly through a mini-narrative or indirectly by association, a secondary female vocalist. Structurally, this woman is mapped onto the role of the background singer, oohing and aahing nonsensically behind the lead's meaningful words (see Corbett 1994:56–67). In other cases, the lead vocals may be sung by a female lead singer who eventually slips into the throes of ecstasy, as is the case on "Love to Love You, Baby."

In all of these songs there is an ambiguity of address: Is the listener being asked to identify with whoever is satisfying the vocalizing woman, or is the listener an outside eavesdropper (the aural equivalent of a voyeur) who "gets off" on the very sound of her voice?[5] Assuming that the ideal listening subject for female orgasm sounds (from the music industry's point of view) is almost certainly male, what happens—as it often must—when the listener is a woman? How does this reorganize the chain of signification? Given a dominant heterosexist perspective, do the cooing sounds of female sexual pleasure serve as a normative model for the "correct" female response to sexual stimulation? Are these sounds part of a disciplinary framework in which supposedly "free" sex vocalizations are ideologically instituted as the acceptable sound of stimulation? Is this a tyranny of ecstasy, teaching women how to sound and men what to try to make women sound like?

The explicit sex sounds used in popular music are clearly often a direct genre reference to mainstream cinematic and videographic pornography. In these forms, the soundtrack during sex scenes will typically relate to the actions that are visually depicted in only the most general way; thus, synchronization is not used as a verification of the "actuality" of the scene. Postsynchronized groans and moans function more as an additional stimulant than as an *effet du reel*. Naturally, this brings us back to issues of "evidence." As Williams says:

> When characters talk their lips often fail to match the sounds spoken, and in the sexual numbers a dubbed-over "disembodied" female voice (saying "oooh" and "aaah") may stand as the most prominent signifier of female pleasure in the absence of other, more visual assurances. Sounds of pleasure [...] seem almost to flout the realist function of anchoring body to image, halfway becoming aural fetishes of the female pleasures we cannot see. (1989:122–23)

Since these female voices in porn film and video are already disembodied from their visual referent, they make a fitting item for purely aural production. Devoid of the usual realist evidentiary role, female sex sounds are free to be used in highly stylized and seemingly antirealist settings. For instance, the samplings of female sex sounds are used in some forms of postdisco dance music in a compulsive-repetitive way. In these contexts, the very same sex sound may be repeated ad infinitum. Though on the surface this appears to be completely nonrealist (as are all mechanistic uses of sample loops and repetitions), at base it still carries deep, "real" connotations about female sexuality. When sampled in this way, these women's voices are hyperrepresentations of female sexuality as out-of-control and excessive. As we have noted, the male orgasm is culturally constructed as terminal and limited, while female sexual pleasure is seen as infinitely renewable and multiple. Like the female orgasm, the technology of sampling is not subject to the generational "exhaustion" of analog technology, but digitally replicates and proliferates the original text. As infinitely repeatable and renewable resources, women's orgasm sounds are thus the perfect item for digital sampling, epitomizing the ecstasy of communication.

Second Station Break

I hear her sigh and I want to buy. I hear her sigh and I want to buy. These words race through my head as I flip through racks and racks of hangers. Euro disco sex pop is piped into my ears, making my head spin. Plastic, leather, latex, scratchy wool, smooth cotton, rubber—a ménagerie of consumable textures. A salesperson comes by and asks if I need help. No, thank you, I'm just looking, I bark. Her sighs grow more intense into shrill, rhythmic, shrieks. The electronic cash register spits and the shoplifting detention device blasts a penetrating alarm. Amid it all, like the filling in pain au chocolat, is the shrieking Euro girl dressed in all the latest fashions.

Cyborgasm

Digital technology has already produced the "first virtual reality sex experience." *Cyborgasm*, produced by Lisa Palac, editor of *Future Sex* magazine, is a compilation of sexual vignettes on compact disc. Modeled largely on *Penthouse* "Forum," the scenarios are almost entirely hetero, including back-seat interracial encounters, light S&M, an orgy, role-playing pedophilia (preceded by a spoken disclaimer by Susie Bright, who assures that the participants are consenting adults), and a science fictionesque dream fantasy about necrophilia. Most are enacted narratives that put the listener in the position of eavesdropper; one utilizes a male voice to describe the sexual event; several include direct address, positioning the listener as part of the diegesis. One thing unites the scenes: almost every cut includes copious female sexual vocalizations.

The prime marketing gimmick that *Cyborgasm* employs is a claim of "Virtual 3-D Audio." Its press release suggests that there are benefits from this technology: "*Cyborgasm* sounds so real you're not just hearing sex, you're having it." Reviewers seem to have bought this virtual line, as evidenced by Jim Walsh's review in *Utne Reader*: "*Cyborgasm* is so in-your-libido vivid, it's like being a fly on the wall of some of your best friends' bedrooms, bathrooms, or back seats. An extremely aroused fly, I might add" (1993:64).

Recorded by Ron Gompertz, whose "Virtual Audio Engineering" earns him auteurlike status in the project, *Cyborgasm* claims (in a sticker on the package) to use "encoding technology developed for virtual reality applications [...] creating a you-are-there listening experience." In fact, the technology for *Cyborgasm* is not new, but utilizes techniques for "audio imaging" long available, primarily side-to-side panning, and foreground/background perspective illusions. These are, at best, somewhat enhanced by the technology of compact disc and "ambisonic" or binaural recording methods (like an audio pop-up book), but they utilize standard studio effects. Especially noteworthy is the gratuitous way that panning is used without reference to activity in the scenarios, suggesting that any unusual, dizzying psychedelic effect will seem three-dimensional. To enhance these effects, the listener is encouraged to experience the disc with headphones, without which the 3-D effect is not heard. "Dim the lights and close your eyes," the instructions read. "Wear our eco-goggles, so you're not distracted by any visual stimuli." Hence, wearing cardboard blinders that come with the CD, one is reminded of the 3-D cinema glasses that were popular in the '50s. But the projection is into the mind of the listener, not onto the screen, or, as the packaging says: "Let your mind go and your body will follow."[6] Obviously, as with most porn, this is a call to masturbate, but without the intrusion of someone else's images into your fantasy. What is significant here is not that this is or isn't a new technology, but that the promotional materials and packaging rely so heavily on claims of technological innovation. Similar claims covered the sleeves of late-'50s record albums, which sought to capitalize on the novelty of stereo and "hi-fidelity." Flamboyant recording tactics (including the use of gratuitous panning) were popular and engendered a profusion of race-car and sound effects LPs. On *Cyborgasm*, it is the promise of new audio technology, the fetish of hi-fidelity that is used to enhance the sex fetish, particularly the fetishization of women's vocalizations. This double sex/tech fetishization, too, has an early precedent. On the back of *Erotica: The Rhythms of Love*, a 1950s LP (Fax Records) that superimposes the sound of squeaking bedsprings and a woman's ecstatic vocalizations over pseudo-Latin drums,[7] an impressive box of text is dedicated to technical data detailing the record's innovative approach. It is claimed to be the "culmination of more than two years of research, utilizing today's most advanced electronic techniques and the talents of sound engineers who have pioneered a host of technical achievements [...] acclaimed by many as a noteworthy landmark in recorded sound."

Whereas *Erotica*'s sex sounds are not supported by narrative justification and explanation, each sex sound in *Cyborgasm* is accounted for in the diegesis of its scenarios. Of these, perhaps the most emblematic of the status of women's voices is "Pink Sweatboxes" by Bunny Buckskin & Carrington McDuffie. This vignette involves a pair of heterosexual female roommates. The first confesses to the other that she gets turned on when she hears her roommate having sex with a male lover through the wall that separates their bedrooms. In particular, the first roommate admits, it is the other woman's wild vocalizations to which she responds. This confession in turn excites the second roommate, predictably leading to a sexual encounter between the two women, undertaken explicitly in an effort to reproduce the coveted sounds. Hence, for the listener, the woman-on-woman scene (not atypical in mainstream hetero-male porn) creates a situation with double-strength female sex sounds. In this narrative, women's voices are used in numerous ways to titillate—as evidence of sexual activity in the room next door, as the "truth" of homoerotic interest, and finally in its traditional role as proof of female pleasure.

Kaja Silverman (1988) suggests that in cinema a compulsive mechanism draws the woman's voice back into the diegesis. In recorded music and aural pornography we find examples that both confirm and contradict this. On one hand, there is the frequent use of sampled sex sounds in current dance music that occurs without reference to a specific narrative. In other cases, like "Je T'Aime," female sex sounds serve as the culmination of the familiar "bringing her to orgasm" story. In either case, the question "What are these sex sounds evidence of?" is left dangling. Without an accompanying image for confirmation, to answer what Rick Altman calls the "sound hermeneutic" (1986:46–47), the question "Is she coming?" can never be answered, "See for yourself." As evidence, the sound of a woman in ecstasy is never quite sufficient for conviction, and the possibility of representing women's sexual pleasure is therefore left ambiguous. But this uncertainty is coupled with an additional representational ambiguity: Are the moans, shrieks, and cries evidence of pleasure or pain?

Third Station Break

Putting the car in drive, I leave the parking lot. It's hot, so I open the window and turn on the radio. They're playing "Love to Love You, Baby," by Donna Summer. I turn up the volume, the car throbs and Donna moans into a slow fade out. As I pull up to a stop light, the disc jockey comes on in a deep, sensuous baritone: "Hey out there in radioland, this is your big daddy deejay." Interrupting his patter, suddenly he begins to groan: "Uh, oh, oh..." A man crossing at the crosswalk shoots me a disgusted look as I turn the radio down. "Mmmmm, aaah..." Instantly, I roll up the windows, despite the heat. "Ahahahahahhhhh." Now, in the privacy of my burning automobile, I'm sweating profusely. "Yes, yes, yes!" With his next outburst, I begin to look at the pedestrians differently. "Ohhhh yes." I'm not thinking about my driving, and accidentally run a stop sign. "Ah, ah...could be enough...gggrrrr, aaah...to satisfy...wheeeeeeeew!"

Crying Dub

Take, as indicative, a late-'70s dub reggae version of Bob Marley's well-known song "No Woman No Cry," called "Crying Dub." Produced by Jamaican dub-pioneer King Tubby, this lyricless dub uses the original song's basic rhythm track, but on top of that and in the place of Marley's original lead vocal it substitutes a woman's voice. This voice is precariously perched between mournful despair and sexual ecstasy. Of course, the title suggests the former, but the quality of the vocalizations themselves suggests the latter. In fact, in almost every example we auditioned, female sexual vocalizations blurred the line that separates a representation of pain from a representation of pleasure, often sounding uncomfortably like screams of torture as much as outbursts of sexual pleasure.

This begs an important question that brings us back to the issue of identification and subjectivity: is the listener assuming a sadistic listening position? Is pleasure, for the listening subject, predicated on a secret (or not so secret) enjoyment of the sound of a woman in pain? Or, on the contrary, is the listener to identify with the vocalizing woman? Does the representation of her pleasure serve as a contradictory place of male-to-female identification similar to Silverman's theory of male masochism (1992:185–213)? Adapting Silverman's theory, the listener (presumed and structured male) may surreptitiously, perversely identify with the woman-as-victim.

Final Station Break

Speed: 7 1/2 inches per second

At the Kinsey Institute, she hands us the sound collection—some vinyl and reel-to-reel tape—and leads us to a small office, closing the door as she leaves. Leather chairs, fancy bookcases, and sexology diplomas. We set up the dusty tape player and push aside "Copulary Vocalization of Chacma Baboons, Gibbons, and Humans" in order to hear the tape of "Sounds during Heterosexual Coitus."

Track: 1/2

Taped to the box is an explanation: "Session I: 6/13/59. Recording begun immediately after intromission. Male face turned partially toward microphone about three feet distant; hence male breathing drowns out female's. Recorded at too-high volume and hence movements on bed give exaggerated noise. Recording ceased after orgasm and when respiration nearly normal."

But it is Session II that interests us: "Eccentric take-up reel causes continual background noise. [...] At orgasm the female gives a series of small cries; subsequently she emits an occasional postorgasmic similar cry. This sort of vocalization is not infrequent in this female."

AV 521

Remnants of sex breath and vocalization leak outside the small office. We try to keep the volume down.

WAD: Timing (From a Hearing 24 Aug. '65 on Wollensak Machine)—

Session II
276 Start Recording
405 Start Heavy Breathing
424
428-33
435-8 female vocalizations
444-7
453-6
472
482-4
509 Verbalization
516 End.

Notes

1. These sounds also appear with some regularity in art music contexts. Hear, for instance, Pierre Schaeffer and Pierre Henry's 1952 piece "Erotica (Symphonie pour un homme seul)," on the musique concrète/electroacoustic collection *Concert Imaginaire* (INA C 1000).
2. Thanks to Keir Keightley for the West and Spivey references. An interesting variation can be heard on Memphis Minnie's 1934 "Moaning The Blues." Here, the (still somewhat eroticized) moan refers not to sexual gratification, but to sadness over the loss of Minnie's man, a subtle combination of pain and pleasure.
3. Contrary to our earlier point, the banning of "Je T'Aime" suggests the possibility that sex sounds are sometimes sufficient evidence to merit strict regulation. A quick listen to "urban contemporary" radio today, however, reveals that these sounds are now publicly acceptable, although words like "pussy," and "dick" are carefully altered for radio play. See Corbett's essay, "Bleep This, Motherf*!#er: The Semiotics of Profanity in Popular Music" (1994:68–73).
4. There are a few converse examples of male sex sounds, including the Buzzcocks's "Orgasm Addict" (1977) and works by audio artist Rune Lindblad and Japanese extremist Gerogerigegege. But the overwhelming majority of examples of bliss noises we uncovered were female.
5. If one of the main issues of certain feminist analyses of mass media has centered on the objectification of the female body, it might be fruitful to ask what becomes of the issue of objectification when the female voice is disembodied?
6. This is a poor paraphrase of Funkadelic's credo "Free Your Mind and Your Ass Will Follow." From *Free Your Mind and Your Ass Will Follow* (Westbound, 1970).
7. Interestingly, these "exotic" drums are accompanied by grunting male musical vocalizations in a stereotyped "ooga-booga" style, implicitly linking the "savage," uncontrollable female sex sounds with the uncivilized "primitive." Only the "civilized," controlled, (presumably) white male protagonist is silent

References

Altman, Rick
1986 "Television/Sound." In *Studies in Entertainment*, edited by Tania Modleski, 39–54. Bloomington: Indiana University Press.

Barthes, Roland
1975 *The Pleasure of the Text.* New York: Hill and Wang.

Bell, Shannon
1991 "Feminist Ejaculations." In *The Hysterical Male: New Feminist Theory*, edited by Arthur and Marilouise Kroker, 155–69. New York: St. Martin's Press.

Corbett, John
1994 *Extended Play: Sounding Off from John Cage to Dr. Funkenstein.* Durham, NC: Duke University Press.

Schaeffer, Pierre, and Pierre Henry
1952 "Erotica (Symphonie pour un homme seul)." *Concert Imaginaire.* LP, INA C 1000.

Silverman, Kaja
1988 *The Acoustic Mirror: The Female Voice in Psychoanalysis and Cinema.* Bloomington: Indiana University Press.
1992 *Male Subjectivity at the Margins.* New York: Routledge.

Sprinkle, Annie
1992 *Sluts and Goddesses.* Video. Produced and directed by Annie Sprinkle and Maria Beatty.

Straayer, Chris
1993 "The Seduction of Boundaries: Feminist Fluidity in Annie Sprinkle's Art/Education/Sex." In *Dirty Looks: Women, Pornography, Power,* edited by Pamela Church Gibson and Roma Gibson, 156–175. London: BFI.

Walsh, Jim
1993 "The New Sexual Revolution: Liberation at last? Or the same old mess?" *Utne Reader* 58 (July/August):59–65.

Williams, Linda
1989 *Hard Core.* Berkeley: University of California Press.
1993 "A Provoking Agent: The Pornography and Performance Art of Annie Sprinkle." In *Dirty Looks: Women, Pornography, Power,* edited by Pamela Church Gibson and Roma Gibson, 176–91. London: BFI.

Developing
A Blind Understanding

A Feminist Revision of Radio Semiotics

Mary Louise Hill

LEAR: What, art mad? A man may see how this world goes with no eyes. Look with thine ears. See how yond justice rails upon yond simple thief. Hark in thine ear: change places and, handy-dandy, which is the justice, which is the thief? (William Shakespeare, King Lear *4.6, lines 150–54)*

When I was originally faced with those basic feminist questions—*Is the gaze male?* and *If so, how can women escape it, manipulate it, or expose it?*—I recalled the answer that Rita in the 1984 film *Educating Rita* gave to the inquiry, "What is the best way to stage Ibsen's *Peer Gynt*?" "On the radio," was her simple reply. Rita's response took into account all of radio's essential semiotic features: being a more time-bound medium than the stage, it is an ideal venue for an epic; being an aural medium, it grants extra signifying power to words, silences, and rhythms. But the one feature that radio also offers a frustrated director of *Peer Gynt* and that drew me to radio's potential as a venue for women's work is the medium's ability to bring life to imaginative landscapes that have been rendered impossible to stage by the realist tradition. As Angela Carter says in the introduction to her collected radio plays, the medium "depends for its effects on the very absence of all the visual apparatuses that sustain the theatrical illusion" (1986:8); because it lacks the concreteness of theatre, including the physical limitations set by the space of a stage, radio would seem to offer the ideal venue for anyone who seeks to transgress what feminists might call "the economy of the gaze."

Just because radio lacks a screen to gaze upon, however, does not mean that residue from the visual economy is also immediately erased. The play I intend to examine here, R.C. Scriven's *A Blind Understanding,* forefronts how radio clings to a realist visual ethic, yet this play also demonstrates how the medium's fundamental *lack* of sight is capable of breaking down the lingering visual apparatus. One of three plays dealing with his blindness, this play reenacts one of Scriven's two unsuccessful glaucoma surgeries.[1] It documents the progress of a man who, at first horrified by the "lack" his blindness consists of, comes to understand that very lack as something with substance.

Luce Irigaray describes most vividly the consequences of possessing a "lack" as an essential characteristic: "[Woman's] sexual organ represents *the horrors of nothing to see.* A defect in this systematics of representation and desire. A 'hole' in its scoptophilic lens" (1985a:26). Irigaray further explains how one who possesses this defect suffers exclusion from the "scene of representation," a fate that both Scriven and radio itself suffered. The change in language used to describe radio reflects how that medium was ultimately shifted out of the primary scene of representation: when television went public in the 1940s, it was spoken of as "radio with added vision," suggesting that radio was, at that time, considered a complete entertainment system. As TV became more accepted, the language shifted to emphasize what radio lacked; it became known as blind, image-less, text-less, and non-visual, thereby reinforcing the television industry's illusion of completeness while relegating radio to a crippled, secondary position (see Drakakis 1981:18–22, 228–29; and Crisell 1986:3–7). Similarly, in becoming blind, Scriven too is forced into a discourse of negativity. He must learn to speak a new language in order to render himself "present" again.

The play begins with a lone clarinet: it rises in a series of dissonant ninths, then descends. At its point of origin it begins to search: up a tritone, down a third, a trill; it is like a hand groping for an unknown light switch, a bird fluttering around a new confinement. Then pause—darkness. A male voice rises out of that pause, announcing: "A knife was driven deliberately into my left eye." Then silence. The same voice persists, deep but slightly unfocused: "I told myself [...] the knife will twist; my eyeball will burst [...]"; his imaginings are interrupted by another voice—a cool, distant, controlled baritone—the surgeon, who instructs him: "From now on you must keep your eye perfectly still [...]." The actual incision is not detailed. Instead, the narrator tries to rationalize the experience, only to continually vacillate with questions such as "How can reason cope with fear?"

A closer examination reveals that the patient fears "the horrors of nothing to see" more than pain itself. But before considering this, I will outline, briefly, the four stages that follow this startling introduction:

1. Three days of "blindness" while his eyes are bandaged, during which he attempts to conjure visual images in order to keep in touch with the "real" world. The primary characters during this stage are the narrator, who acknowledges his fate is in the hands of another man, and the doctor—the other man—whose voice always seems to come from somewhere slightly above and to the left of the narrator's. During this time, that narrator also accustoms himself to the sounds of his environment.
2. The unbandaging and restoration of sight, and the accompanying knowledge that it is temporary. The narrator still feels his fate is in another man's hands. He leaves the hospital with the intent of seeing everything as if it were his last sight, and of storing all those visions in his memory. He finds himself in conflict between a *carpe diem* attitude and one that looks inward, and he attempts to develop his spirituality. This conflict is embodied in the characters DAY (spoken by a man, another baritone) and NIGHT (a woman's voice, mezzo-soprano).
3. The attempt to see clearly and rationally, dominated by DAY's guidance. The narrator is not content with this, as he wants to develop a poetic vision.
4. The development of poetic vision, attained by looking with the "inward eye." NIGHT emerges as a guide that the narrator at first rejects, largely out of fear. Gradually, he accepts the darkness, the unknown, and the ambiguity within it.

Such a cursory review might suggest that *A Blind Understanding* is a simple story, but its structure betrays a more complex sound experience. As with many radio dramas, this play contains two aural *zones* in which the "action" occurs, sometimes nearly simultaneously. The first, "public" zone, established in the surgery, is where the narrator interacts with other people, especially the doctor. The second zone, on the other hand, includes the narrator's "personal" reflections, fears, and images, including his symbolic characters of NIGHT and DAY.

The presence of these two concurrent zones complicates any generalizations about radio as a linear medium dictated by narrative rules. Though it is true that spoken words take more responsibility than physical imitation in radio, language acts in counterpoint, often associatively, rather than consecutively. Words act in conjunction with sounds and silence to produce a polyphony similar to that attained by a symphony.

The public zone of *A Blind Understanding* implements those sound and language patterns most commonly found in "realist" radio. Dominating approximately the first third of the play, this zone provides a bridge to the visual world that the narrator desperately wants to remain a part of. Therefore, he selects and fixes on sound signs that will help him produce a picture of the real world. Yet the type of sign available in the blind environment of radio drastically truncates the link to the "real" world, even more so than theatre signs do.

In theatre, the primary sign system is comprised of *icons*, actual objects which—in realism—rely upon their *similarity* to the absent object for meaning.[2] The theatre audience experiences a multiplicity of icons in the mise-en-scène (see Pavis 1992:31), and their interpretation depends on the audience's ability to read on at least two levels. First, theatre signs are read according to the accepted rules of the stage and the particular rules of the specific text or performance. This theatrical reading is supported by a secondary, cultural reading (see Alter 1990). If strong enough, the cultural reading might transform how the stage sign is read. To illustrate this, I think of Louis Malle's 1995 film *Vanya on Forty-Second Street*. The text calls for the samovar, but the character brings forth a glass pitcher. Especially since the action of this filmed play occurs on an undressed, crumbling stage, an audience member is more likely to connect a personal, culturally inscribed meaning to the actual object, and humor arises from the contemporary, utilitarian meaning and how it conflicts with or deflates the "stage meaning" of the samovar.

Radio signifiers likewise attempt to mask an absence, but they do not replace that absence with a thing and thereby produce an illusion of presence. The radio sign works indexically—causally, or as an indication—rather than iconically. A tapping noise, for instance, is not *like* a tree branch hitting against a window pane; it is like the sound that would result from that movement. Therefore, the absence is compounded by the uncertain source of the sound. Scriven's public zone adapts the same strategy most radio practitioners have implemented to accommodate this uncertainty. Implementing a primary system of radio reference, Scriven isolates and amplifies such sound signs as footsteps, chiming bells, and traffic sounds. Along with the narrator, the listener quickly adopts and clings to these signs' "stock" meanings, which are then reinforced by a second, cultural message. For instance, the listener's personal experiences of a hospital facilitate an understanding of Scriven's narrator's own anxiety. An understanding of Scriven's trauma over losing his sight is possible simply because we all live in a culture that privileges sight.

The heavy reliance on the cultural connotations of radio signs alerts us to the indexes' inherent incompleteness, a characteristic that puts a natural division between the radio world and the "real world" and produces a field of potential ambiguity. This potential ambiguity of uncontextualized sounds has been controlled by adapting repetition and linguistic "signposting," a strategy eagerly utilized by the narrator in Scriven's public zone. Heavy footsteps he labels "nurse";

a light busy step he calls a child. Once again, this textual pointing serves to bring the narrator, and simultaneously the radio listener, closer to the visual world.

Andrew Crisell accurately connects this "textual pointing" to Roland Barthes's theory of captions for photographs: "Visual images, [Barthes] argues, are polysemous. But so are sounds. Hence words help *fix* the floating chain of signifieds in such a way as to counter the terror of uncertain signs" (1986:51).[3] This particular quote is very telling when applied to the radio event, for without the textual pointers, the listener would be confronted with a floating chain of aural signifiers and no clues as to their signifieds. The terror, then, that textual pointing counteracts is one of a coherent visual image splintered; ultimately, textual pointing maintains, in a very fragile environment, the illusion of a unified subject.

Such an illusion of unity is a substantial part of the ideology Scriven's captions seek to perpetuate. Ordinarily, as a man in this world, Scriven would assume the role of the healthy and therefore privileged, but he must now accept the fact that he is incomplete. His statement, "Whether or not I see the light of day must depend on other men," marks his demise as a total man. In fact, we experience his new role from the outset when, prone on the operating table, he must unflinchingly submit to the surgeon's descending knife. Although the surgeon says, "I do not wish to dominate, as much as you do not wish to be dominated," both happen. One man emerges empowered; the other, emasculated. Trapped within the blindness of the medium as well as that of the narrator, the listener also suffers a physical loss; together the narrator and listener must develop another way to maneuver the sight-less space.

This is where the personal zone of *A Blind Understanding* begins to assume more importance. In the wake of this symbolic castration, Scriven turns more and more inward, and his emerging personal voice assumes a variety of characteristics. Most notable is the timbre of the internal voice itself, especially when compared to the doctor's. Both characters are portrayed by baritones,[4] but whereas the doctor's voice is crisp and centered (a listener detects that Barthesian *grain*—"The body in the singing voice" [1985:276; 1977:188]), the narrator's voice is unfocused, breathy, and heady. In fact, it seems to be losing its body. Especially in sections when he elaborates upon his fear of darkness, one might even conclude that he is suffering hysteria, that malady that Freud so often ascribed to women. But no matter what the diagnosis (or its prognosis), the fact remains that by the time the narrator leaves the hospital, the personal zone has become the primary zone of this play.

The shift toward a primarily internal drama constitutes a departure from the realist influence of the public zone and an adaption of a purely "radio" perspective. Recognizing that he must submit to the primacy of sound in his blind world, the narrator begins developing his own kind of noise. In a fashion reminiscent of a Beckettian narrator, his constant stream of words attempts to fend off the encroaching darkness and silence and to establish a new type of bodily presence, as well as a new ideology.

The first strategy he adapts is poetic structure. His lines become recognizably iambic, often with regular, mappable rhymes. By implementing this, he produces an awareness of an absent textual object as well as an everpresent rhythmic object. In fact, after first listening to this play, I remembered the rhythmic pattern more easily than the words themselves and thereby realized how easily radio can transform words to mere sound objects, floating signifiers freed from their usual referents.

Those few isolated words that persisted in my memory were symbols and allusions. In an attempt to understand the darkness he fears, and to communicate why he fears it, Scriven turns to established literary figures and tropes: to Hamlet, to Lear, and to Louis MacNeice's classic radio play "The Dark Tower."

Such literary allusion works in a way similar to metaphor, juxtaposing similar literary moments in order to develop an idea. In declaring that "metaphor occurs at the precise point at which sense emerges from non-sense," Lacan reminds us that metaphor produces a signifying situation in which signifiers refer to yet other signifiers and the signified continually slips beyond our reach (1977:158). Similarly, literary allusion draws us back to another written moment, one that may have yet other allusions within it. The greatest difference between allusion and metaphor is that allusion grants the writer a history. Tapped into that history, he can claim authority as well as a larger narrative frame.

Through his adaptation of poetic allusion and structure, Scriven calls upon another established patriarchal paradigm to help him regain his status as a coherent subject. But that wholeness is continually undercut by the deferred referent that lies just beneath the surface of allusion. In addition, the symbolic conflict between DAY and NIGHT that occurs within his imagination provides ample contradiction to his poetic edifice. DAY's relationship with NIGHT fluctuates; first she is his enemy, then his sister. As the narrator says, "Against reason, I am swayed by her," and his reason falters whenever her throaty voice arises from the silence, promising to live with him "closer than a lover."

Clearly these symbolic characters represent the simple binaries of reason and passion, monotheism and atheism, society and antisociety. Allied with these binaries, it is not surprising that DAY be characterized as male, NIGHT as female. Speaking out of a generation that enforced these binaries, Scriven's reluctance to depart from the visual betrays itself fully when he characterizes the blind radio space as female. For him, and for many like him, submission to radio's blindness equals a feminizing subjugation, which includes accepting all of the social stigmas of being woman.

Still, he does finally submit, at the very end, when he denies God and the flesh and quixotically declares, "I can see! I can see!" before letting silence overpower him and the listener. These last minutes produce a total breakdown of order for the listener as the narrator's voice lifts into a higher register, and he admits to almost total uncertainty.

In these last moments, as the narrator accepts the uncertainty and ambiguity of being without sight, he places a large part of the responsibility for producing meaning on the listener's shoulders. Certainly, radio's performance "space" is more in the listener's mind than it is in the studio; terms such as "the theatre of the mind" (see Esslin 1980:171–87) produce images of a dusty, empty stage located somewhere between the ears, just waiting for the performers, sets, and props to be beamed in from on high. If this were the case, I could discuss radio performance in terms of mise-en-scène, but, as I have already established, radio does not possess the essential concreteness of the theatre space. Further, as Martin Esslin points out, "It is precisely the nature of the radio medium that makes possible the fusion of an external dramatic action [...] with its refraction and distortion in the mirror of a wholly subjective experience" (1986:365), reminding us that radio obliterates at least two other elements of realist theatre: the gulf between the stage and the auditorium, and the "fourth wall." With this gone, the listener joins the speaker and shares in his experiences. How the listener assigns meaning or objecthood to each signifier is totally up to that listener's own linguistic, cultural, and social circumstances.

If anything, the experience of listening to radio is closer to that of translating, a mental activity that Patrice Pavis has described as *mise en jeu* (1992:136–59). At the heart of the translation, mise en jeu is an *exchange* from the source text to the target language "effected by comparing and trying out word and object presentations in the two languages and cultures and in adjusting the language-body of the two systems accordingly" (148). This comparing and trying out, according to Pavis, occurs in an unconscious, preverbal situation inside the translator's imagination, and it involves accommodating not only words but also their sound picture, as well as their cultural and physical implications. In a similar fashion, a radio listener encounters the words and sounds on the imaginative terrain of the mind. That listener, too, compares and tries out meanings, images, physical and emotional responses within the imagination, in order to produce an image or response. Also similar to translation, these images or responses are not fixed: they develop and change as the experience continues. But unlike translation, radio does not demand concretization; its signs remain in a preverbal state, producing a continual kaleidoscopic juggling of words, sounds, and images—perhaps more literally, a mise en jeu.

But there is more at play in radio's mise en jeu than words and sounds. The two zones that Scriven's work asks both the listener and narrator to accommodate point to how radio enforces a split, if not multiple, subjectivity. What is impressive about Scriven's play is that in it we actually witness the severing of the subject when the surgeon (assuming an active subject role despite his off-center vocal presence) penetrates the narrator's eye. The narrator, though acoustically located in the center, is the object he acts upon. As long as he is in contact with the surgeon, the narrator maintains an object position that he, with his binary-prone mind, might normally assign to a woman. He is repeatedly examined, bandaged, unbandaged, and assured that he must undergo surgery (penetration) again.

Therefore, that first surgical gesture severs the narrator from his own illusion of coherent subjectivity at the same time that it objectifies him. His subsequent verbalizations constitute attempts to reclaim active subjecthood. A review of the language of Lacan's mirror stage theory illuminates this process. "The *mirror stage*," Lacan claims,

> is a drama whose internal thrust is precipitated from insufficiency to anticipation—and which manufactures for the subject, caught up in the lure of spatial identification, the succession of phantasies that extends from a fragmented body-image to a form of its totality that I shall call orthopaedic—and, lastly, to the assumption of the armour of an alienating identity [...]. (1977:4)

The key words in Lacan's description of this stage are "drama" and "phantasies," for they indicate how this entire process is a fictional act: this "drama" consists of the stories the individual tells himself about himself. At the socialization stage, the stories that coincide with the accepted cultural story are retained as the armour of an alienating identity; those that do not are sublimated. But, significantly, that split between the social self and the unconscious self *remains*, though the unconscious is repressed, back in the preverbal stage. With this in mind, we understand that what has essentially happened to Scriven's narrator in the operation scene is that, in losing his sight and his privileged subject position, he is forced back into the split, and he must begin again the process of fictionalizing.

The two zones the play occupies coincide precisely with the two levels of the Lacanian split—the social and the unconscious. In addition, the social narrator suffers another split, which corresponds with Lacan's assessment of how woman socializes herself. The doctor—the coherent male subject—acts as a "relay whereby the woman becomes this Other for herself as she is this Other for him" (1982:93). She must, therefore, fictionalize the male subject position in order to construct her own, just as Scriven must try to see what the doctor sees, an act at which he realizes—as a woman must realize—he will continually fail because of the doctor's specialized knowledge about his illness. So he must accept his failure and construct a new language for himself, and this is what emerges in the personal zone.

Sue-Ellen Case calls this a further fracturing, or a second split in the female subject, which produces a "split-subject-in-discourse," a woman who "cannot appear as a single, whole, continuous subject [...] because she senses that [man's] story is not her story" (1989:131). A *split* discourse would include both a running commentary on the woman as object-to-be-viewed and an account of her efforts to construct her own subjectivity. To this she adds the idea of the "metonymically displaced" subject, in which "the subject further fractures into multiple displacements across the stage and is transmutable during the course of the play" (134). Notably, her examples—which include Hélène Cixous's 1976 *The Portrait of Dora* and Marguerite Duras's 1972 *India Song*—implement voice-over, monolog, and disembodiment to indicate the displaced female subjectivities. In these examples, as in radio, the voice itself is the metonymy for the absent body, the object for which the displaced female voices express a desire.

Scriven also displaces aspects of himself in the interplay between his inner voices DAY and NIGHT. NIGHT, in particular, demonstrates how his multiple subjectivities include at least one previously repressed bodily fiction, which is marked by a *female* voice. Together with DAY, the narrator, and the doctor, these voices create a chorus circling around his absent body, one which actually gains substance in the listener. Left to refictionalize a coherent subject both for the narrator and for herself, this listener becomes, to use Duras's words, "an echo chamber. Passing through that space, the voices should sound, to the [listener], like [her] own 'internal rending' voice" (in Case 1989:136). In negotiating these displacements, the listener embodies all of the speaker's absent desires.

What that listener essentially "embodies," though, is the mise en jeu itself, a state of imaginative flux and change wherein nothing is really fixed. Irigaray might have been thinking of something similar when she described how woman should battle against the oppression of words:

> Turn everything upside down, inside out [...]. Rack it with radical convulsions. Insist also and deliberately upon those *blanks* in discourse which recall the places of her exclusion [...]. *Overthrow syntax* by suspending its eternally teleological order, by snipping the wires, cutting the current, breaking the circuits, switching the connections, by modifying continuity, alternation, frequency, intensity. Make it impossible for a while to predict whence, whither, when, how, why [...]. (1985b:142–43)

Here, Irigaray seeks what Scriven's play leads us to: a place where, freed from reasonable words, realist images, even accepted subjectivities, a listener is left to play with whatever words, images, and self-fictions lie within her own repressed unconscious. Within radio's mise en jeu, in fact, the listener does not even have to fix on one subjectivity; for as long as one is in the radio state, the freedom to play continues.

What is most curious about Irigaray's passage is her use of an electrical metaphor to give shape to an oppressive language system. In the context of my discussion, a very specific concrete image emerges in response to her metaphor: a seemingly benign little radio. In my case, it's the one that hides behind a pile of papers on my kitchen counter.

Feminism makes us aware that that object is not so benign. Radio *is* the physical apparatus Irigaray calls for to snip the wires of the teleological order; its gaping silent mouth eagerly waits to give body to all of the needs of a discourse characterized by what it lacks.

Notes

1. As a radio writer, Scriven himself is quite an anomaly. He suffered from impaired hearing (he was deaf in one ear) from the age of seven. During his forties, he began losing his sight due to glaucoma, and was completely blind by his early fifties. Yet, as his son Marc informs me, he continued to write both radio plays and poetry until he died in 1985, at the age of 78. Several of his radio plays focus on that blindness, including *A Single Taper, Seasons of the Blind,* and *A Blind Understanding.* Broadcast on BBC 4 on 4 October 1977, *A Blind Understanding* is a version of the earlier *A Single Taper* (1947). Since the texts of Scriven's work are out of print and unavailable as of this writing, my details on Scriven and his work are based upon correspondence with his son Marc and my own inscription of tapes of his plays' original performances, heard at the National Sound Archive of the British Library.
2. See Keir Elam (1980:21–32) for more on iconicity in the theatre. Also, see Marvin Carlson's "The Iconic Stage" (1990:75–91).
3. See also Roland Barthes's "Rhetoric of the Image" in both *Image - Music - Text* (1977:32–51) and *The Responsibility of Forms* (1985:21–40).
4. David March performs Scriven's surgeon; Steven Murray, the narrator.

References

Alter, Jean
1990 *A Sociosemiotic Theory of Theatre.* Philadelphia: University of Pennsylvania Press.

Barthes, Roland
1977 *Image - Music - Text.* Translated by Stephen Heath. New York: Hill & Wang.
1985 *The Responsibility of Forms.* Translated by Richard Howard. New York: Hill & Wang.

Carlson, Marvin
1990 *Theatre Semiotics: Signs of Life.* Bloomington: Indiana University Press.

Carter, Angela
1986 *Come unto These Yellow Sands.* Newcastle upon Tyne: Bloodaxe Books.

Case, Sue-Ellen
1989 "From Split Subject to Split Britches." In *Feminine Focus: The New Woman Playwrights,* edited by Enoch Brater, 126–46. New York: Oxford University Press.

Crisell, Andrew
1986 *Understanding Radio.* London: Methuen.

Drakakis, John, ed.
1981 *British Radio Drama.* Cambridge: Cambridge University Press.

Elam, Keir
1980 *The Semiotics of Theatre and Drama.* London: Methuen.

Esslin, Martin
1980 "The Mind As Stage." In *Mediations: Essays on Brecht, Beckett, and the Media.* Baton Rouge: Louisiana State University.
1986 "Samuel Beckett and the Art of Radio." In *On Beckett: Essays and Criticism,* edited by S.E. Gontarski, 360–84. New York: Grove Press.

Irigaray, Luce
1985a *The Sex Which Is Not One.* Translated by Catherine Porter. Ithaca, NY: Cornell University Press.
1985b *Speculum of the Other Woman.* Translated by Gillian C. Gill. Ithaca, NY: Cornell University Press.

Lacan, Jacques
1977 *Écrits: A Selection.* Translated by Alan Sheridan. New York: Norton.
1982 *Feminine Sexuality.* Translated by Jacqueline Rose. New York: Norton.

Malle, Louis

1995 *Vanya on Forty-Second Street.* With Julianne Moore, George Gaynes, Andre Gregory, and Wallace Shawn. Columbia Pictures.

Pavis, Patrice

1992 *Theatre at the Crossroads of Culture.* Edited and translated by Loren Kruger. London: Routledge.

Russell, Willy

1984 *Educating Rita.* Directed by Lewis Gilbert. With Michael Caine and Julie Walters. RCA/Columbia.

Scriven, R.C.

1977 *A Blind Understanding.* Charles LeFaux, producer. British Broadcasting Company, Radio 4, 4 October.

Interpolation and Interpellation

Fred Moten

Imagine an incessant listening by which one might be engaged or called. Such a listening might provide great pleasure and, in so doing, produce consternation and anxious questioning about the nature of such pleasure. Those questions might concern the psycho-political effects or politico-economic grounds of the submission of oneself to such pleasure. But, in the end, both the fact and the depth of the questioning that is produced by checking over and over again, say, Bach's *Mass in B Minor*[1] seems always to amount to something that's all good.

Over the past few years I've been caught up in an obsessive relationship with a song called "Ghetto Supastar" (1998), performed by Pras, Ol' Dirty Bastard, and Mya. The song was produced by Pras and Wyclef Jean and contains what is referred to in the liner notes as an "interpolation" of Barry Gibbs's "Islands in the Stream" (1982). Let me see if I can draw you in as well to the complex interplay of pleasure and questioning that is produced by "Ghetto Supastar." Now there are those who would never admit the possibility that such an object could produce such interplay. For some, this would indicate the object's fundamental lack of aesthetic value. For others who would champion the devaluation of the aesthetic, "Ghetto Supastar" might be valorized precisely because it has no value. I want to argue here against the possibility of both formulations.

Let me repeat, then, that the pleasure I derive from "Ghetto Supastar" raises questions. How could one derive pleasure from such a thing? And, if you've got some inkling about the first question, what's it mean to derive such pleasure? Is such pleasure what Theodor Adorno would disparage as "culinary" (1982:273) or temporary, an effect of an evacuation of reason that's all bound up to a certain giving up of, which is to say, giving oneself up to, the body and its base or basic (or bassic) functions? Is this a kind of pleasure that's tied too much—too much a function of—the hook, of being hooked to or by the addictive repetition of a catchy tune? Or, more drastically, does this song reveal to us that which requires Louis Althusser to claim that aesthetic judgment in general is "no more than a branch of taste, i.e., of gastronomy" (1971:229). Either way, what I'm trying to do here is think about this song's flavor and my pleasure in it within the context of the theme of utopia and, both more broadly and more specifically, the theme of (the relationship between) scholarship and commitment. I want to think these themes in relation to a desire for music that is properly and unanxiously political. That which Cedric Robinson calls, in his indispensable *Black Marxism* (1983), "the black radical tradition" has been acting on and out and theorizing that desire for a long, long time.[2]

The theme of scholarship and commitment in its relation to the political desire for music has a reanimative, *quickening* function. It brings life. Noise. It offers up a bit of anima/soul/breath and therefore serves to air out various venues that had been overwhelmed by the scent of a kind of putrefaction, the smell of death that hovers over even those spaces where folks are talking about resistance or hybridity or citizenship or whatever while believing or being driven by a belief that the times we live in or the modes of thinking that are now prevalent are, to use Judith Butler's term, "post-liberatory" (1997:17–18).

It might be romantic, but for me, music—and here I include the sound or spirit of the refrain of scholarship's relation to commitment—messes up the very idea of the post-liberatory. But the point, here, is that it's not enough to say that we need music, or even to imagine for oneself a relation to the various contexts of our various, often embarrassed, discussions of freedom that would invoke Q-Tip, formerly of A Tribe Called Quest, when he says that "the job of resurrectors is to wake up the dead" (1991). This leads us to some more questions: first of all, there is the question of why anyone would want to revive such contexts; second is the question of the incorporation of the music into the larger culture, into the culture industry, and thereby, into an ensemble of relations of cultural production that also now determine the corporatized and industrialized production of academic stuff, which is to say not only what is called academic knowledge but also what are called academics; third is the question of the incorporation of "Ghetto Supastar" into the soundtrack of Warren Beatty's 1998 film *Bulworth*. This question, in its turn, raises others: that of the film's incorporation of the black radical tradition and its sounds; that of the American left's ambivalent interpellation of that tradition in general; and its insight—at once primitivist and progressive—concerning that tradition's special, vexed, complex, hopeful relation to aesthetic, political, and libidinal (enslavement and) freedom. This, along with a valorization of a kind of hybridity that ought to make anyone who has ever valorized hybridity pause, is what the film both represents and enacts. But I'll let this slide too, though I would point out that the last ensemble of questions requires at least some thought regarding the fetish character of Halle Berry and her secret. What I will try to get to is bound up with the radical impossibility and undesirability of detaching the fetish character of the commodity from the commodity.

The main thing, in the end, is to think about what the foreclosure of music has wrought where music is understood not only as a mode of organization but, more fundamentally, as phonic substance, phonic materiality irreducible to any interpretation but antithetical to any assertion of the absence of content. This could be about the utopian function of dropping science where dropping signifies a critical, if appositional, engagement and dissemination rather than a mere and misguided dismissal. Obviously this interplay of science and utopia has to do with Marx, with the divided Marx that Althusser produces (without discovering), with the Marx that is beyond Marx, according to Antonio Negri (1991), or the Marx that is before Marx. This last Marx, the one that is before Marx, is the one in which I'm most interested; this is the Marx that is anticipated in and by the black radical tradition. This essay just responds to a Marxian interpellative call that was itself anticipated by the black radical tradition, always already cut and augmented by an anticipatory interpolation. "Ghetto Supastar" extends that tradition by exemplifying its formal operations, even if on what some would think of as modest terrain, terrain awaiting the exaltation that life-giving and anticipatory revision makes possible. This is, in other words, what Pras and Maya and Ol' Dirty Bastard do to and for "Islands in the Stream," which you might remember in the version performed by Kenny Rogers and Dolly Parton, a tune that did not fully exist—or, more precisely, was not fully alive—until Pras killed it. In this sense,

Pras does to and for that tune what Trane does to and for Rodgers and Hammerstein's "My Favorite Things" over and over again. The black radical tradition has numerous other examples of such anticipatory interpolations, revisions of the original that give it birth, while evading, as Nathaniel Mackey (1986) might say, the natal occasion.

This is all to say that the cut calls one to think about the ways the trajectory of black performances, which is to say black history, constitutes a real problem and a real chance for the theory of history as such. One of that trajectory's implications, if it is set to work in and on such theory, is that those manifestations of the future in the degraded present that C.L.R. James describes both by way of and against Marx (1977), can never be understood simply as illusory. The knowledge of the future in the present is bound up with what is given in that which Marx could only subjunctively imagine: the commodity who speaks. Here, again, is the relevant and very well-known passage from volume one of *Capital*, at the end of the chapter on "The Commodity," at the end of the section called "The Fetishism of the Commodity and Its Secret," so that we can check the reasons for Marx's "impossible" example:

> As the commodity-form is the most general and the most undeveloped form of bourgeois production, it makes its appearance at an early date, though not in the same predominant and therefore characteristic manner as nowadays. Hence its fetish character is still relatively easy to penetrate. But when we come to more concrete forms, even this appearance of simplicity vanishes. Where did the illusions of the Monetary System come from? The adherents of the Monetary System did not see gold and silver as representing money as a social relation of production, but in the form of natural objects with peculiar social properties. And what of modern political economy, which looks down so disdainfully on the Monetary System? Does not its fetishism become quite palpable when it deals with capital? How long is it since the disappearance of the Physiocratic illusion that ground rent grows out of the soil, not out of society?
>
> But, to avoid anticipating, we will content ourselves here with one more example relating to the commodity-form itself. If commodities could speak they would say this: our use-value may interest men, but it does not belong to us as objects. What does belong to us as objects, however, is our value. Our own intercourse as commodities proves it. We relate to each other merely as exchange-values. Now listen how those commodities speak through the mouth of the economist:
>
> Value (i.e., exchange-value) is a property of things, riches (i.e., use value) of man. Value in this sense, necessarily implies exchanges, riches do not.
>
> Riches (use-value) are the attribute of man, value is the attribute of commodities. A man or a community is rich, a pearl or a diamond is valuable . . . A pearl or a diamond is valuable as a pearl or diamond.
>
> So far no chemist has ever discovered exchange-value either in a pearl or a diamond. The economists who have discovered this chemical substance, and who lay special claim to critical acumen, nevertheless find that the use-value of material objects belongs to them independently of their material properties, while their value, on the other hand, forms a part of them as objects. What confirms them in this view is the peculiar circumstance that the use-value of a thing is realized without exchange, i.e., in a social process. Who would not call to mind at this point the advice given by the good Dogberry to the night watchman Seacoal?

> "To be a well-favoured man is the gift of fortune; but reading and writing comes by nature." (Marx [1867] 1990:176–77)

The example is given in order to avoid anticipation but the example works in such a way as to establish the impossibility of such avoidance. Indeed, the example, in her reality, in the materiality of her speech as breath and sound, anticipates Marx. This anticipation is the detour that anticipates the one we now inhabit. It's the anticipatory detour that animates "Ghetto Supastar." I want to try to recover this sound that is, itself, anything but originary. This sonic event was already a recording, just as our access to it is made possible only by way of recordings. We move within a series of phonographic anticipations, messages encrypted, sent, and sending on (lower?) frequencies that Marx tunes to accidentally, for effect, without the necessary preparation. This absence of preparation or foresight in Marx, an anticipatory refusal to anticipate, is, though, the condition of the possibility of a richly augmented encounter with the chain of messages the speech or sound of the commodity carries. Or, more precisely, the intensity and density of what could be thought here as his alternative modes of preparation, make possible a whole other experience of the sound of the event of the commodity's speech, a whole other experience of this event/music.

Moving, then, in the critical remixing of nonconvergent tracks, modes of preparation, traditions, we can think how the commodity who speaks, in speaking, in the sound—the inspirited materiality—of that speech, constitutes a kind of temporal warp that disrupts and augments not only Marx but the mode of subjectivity that the ultimate object of his critique, capital, both allows and disallows. All this is to say that I want to move toward the secret Marx revealed by way of the music he subjunctively mutes. Such aurality is, in fact, what Marx called the "sensuous outburst of [our] essential activity" ([1844] 1975:356). It is a passion wherein, he might say, "the *senses* have therefore become *theoreticians* in their immediate practice" ([1844] 1975: 352). The commodity who speaks—and in speaking, sounds—embodies the critique of value, of private property, of the sign. Such embodiment is also all bound up with the critique of reading and writing—oft conceived not only by clowns but by intellectuals as the natural attributes of whoever would hope to be known as Man—that Pras et al. instantiate.

In the meantime, every approach to Marx's example must move through the ongoing event that anticipates it, the real event of the commodity's speech, itself, broken by the irreducible materiality of the commodity's sound. Imagine that there's a recording of the (real) example that anticipates the (impossible) example. Imagine that recording as the graphic reproduction of a scene of instruction, one always already cut by its own repression. Imagine what cuts and anticipates Marx, remembering that the object resists, the commodity sings or shrieks, the audience participates. Then you can say that Marx is prodigal, belated; that in his very formulations regarding Man's arrival at his essence, he fails to come to himself, to come upon himself, to invent himself anew. This failure is at least in part the failure to reveal a kind of internal secret—this in the one whose entire project is characterized by an attunement to the revealed secret. What remains secret in Marx could be thought of as or in terms of race or sex, of the differences these terms mark and reify. But we can also say that the unrevealed secret is a certain recrudescence of the private. He can point to but not be communist and what the dispropriative event and its music have to do with communism is at stake here. What's the revolutionary force of the sensuality that emerges from the sonic event Marx subjunctively produces without sensually discovering? To ask this is to think of what's at stake in Pras's music: the universalization or socialization of the surplus, the (re)generative force of a venerable phonic propulsion, the ontological and historical priority of resistance to power and objection to subjection, the old-new thing, the freedom drive.

You could think about this drive in many ways, but for me it always goes back to the fact that the black radical tradition is first manifest in and given as the unimaginable speech of the commodity, in the irreducible phonic substance and the irreducible kinetic materiality that instantiate, accompany, and disrupt that speech and its interpretability. That sound, for instance, is often given as the response to demands for recognition that emerge as interpellative calls or other such passionate utterances. That sound often manifests itself as the appositional or arrhythmic cutting of that passion, from within passion. Here the screams, for instance, of Frederick Douglass's Aunt Hester in her passion, might disturbingly but generatively be thought of—by way of the constitutive force they exert on the discourse regarding the music of the black radical tradition[3]—in their aesthetic and political relation to the sounds of Bach's *St. Matthew's Passion*. But that relation would have to be established by way of an immersion or Ellisonian lingering in the music that I can only point to today. Such music cuts the masterful force of certain passionate attachments with the serrated edge of a radically exterior, radically sentimental lyricism. Douglass is reanimated by such music as Pras makes. Such music can move in the novel as well.

This brings me to another version of "Ghetto Supastar," a novel co-written by Pras and kris ex. The novel begins by anticipating its end. It tells the story of Diamond St. James, a young rapper struggling to escape the dangers of a (stereo)typical urban nightmare by entering into the even more fearsome world of the culture industry. Diamond's struggle is to be in and not of the latter, to be of and not in the former. His music would both instantiate and represent that struggle. The opening of the novel anticipates its end by showing us a mature and successful Diamond, fending off the interpellative telephone calls of his publicist and of his old partner in street crime, the now incarcerated Michael. Michael's call literally interrupts, via call waiting, that of the industry given in the form of a call from Diamond's agent, knowingly figured as the one who endangers, by way of a kind of protection, Diamond's agency. You could say, then, that the novel is all about the vexed possibilities of resisting interpellation, a possibility given in musical interpolation. But Althusser (1971a) makes sure to let you know that interpellation is, in essence, more fearsome than these initial examples. The interpellative call is exemplified by the call or sound of the police rather than that of the publicist or old running buddy. The thing is, Pras figures that more fundamental and dangerous figure in the novel as well. Check it out:

> He was two blocks away when he noticed a beat cop eyeing him knowingly.
> Diamond didn't know what to think. He'd been through so much, done so much, seen so much in the past week, that he knew he was under arrest for something. It didn't matter what it was—he was sure that he was guilty. It was a long time coming, and he knew the rules—he wouldn't turn on Michael or Mr. B for all the amnesty in the world.
> "Hey," the cop called, moving closer with a smile. "What's up?"
> Diamond just looked at him. He knew the tricks, the traps, and the runnings. His mouth stayed shut.
> The officer grinned widely. "You don't remember me, do you?"
> Diamond peered underneath the man's cap.
> "St. James—no Walkman's in the hall," he bellowed with a laugh.
> Diamond smiled. It was Nixon—he used to be a security guard when Diamond was in high school.
> "What's up man" Nixon smiled. "What you been up to? How's the rapping thing? I hear you're still doing it."

> "Yeah, yeah, no doubt," Diamond said anxiously. He was never more than passing acquaintances with the man when he was in high school, and now, even in parts removed from his home turf, he did not want to be seen cavorting with the police.
>
> "You ever thought about getting down with the force?" Nixon asked with a recruiting smile. "It pays pretty well. You know," he said with secrecy, "we need more brothers in blue."
>
> "Yeah," Diamond said. He needed to get the conversation over with as soon as possible.
>
> "I'm fresh out the academy," beamed Nixon, proudly. "They'll be testing again in a few weeks. You don't want to miss it."
>
> "Yeah, yeah."
>
> "All you have to do is go down to a precinct and pick up an application—you have your diploma, right?"
>
> "Nah," Diamond lied, hoping it would be deter the man and send him on his way.
>
> "Well, you gotta get your G.E.D., brother," Nixon advised. "And sanitation is testing next Tuesday. It may be too late for that one—but my sister works there, she could help you out if you really want it. Let me get your number and I'll call you with all the info."
>
> Diamond rattled off any seven numbers. The cop asked him to repeat the numbers four different ways, obviously part of his training to discern when those damned detainees were trying to pull a fast one. Had he not so much practice memorizing his rhymes he would have faltered.
>
> "Alright, brother," Nixon said, slapping palms with the boy off beat, proud that he was able to merge his academy training, street savvy, and inside connects to the hood in one triumphant moment of community policing. He would run a search on the boy when he got back to the station house that night. Hell, he'd make sergeant sooner than anyone thought. (Michel 1999:149–51)

I can't say too much here about this scene but I do wish to point out a couple of things: one is that Diamond comes upon the policeman and his gaze, rather than being surprised by it from behind in the now classical Althusserian scene. Part of what this gets to for me is the impossibility of a certain kind of surprise as such scenes are transposed into a different venue and recast with different prospective subjects. It takes a special kind of subject-in-waiting to be surprised by the presence of the police or, more problematically, to respond to that surprising hail in a way that betrays what Butler calls a "passionate attachment" to the law. (1997:6–10). Happily, this special kind of subject-in-waiting is not the universal model. Instead, we've got Diamond, the sentient, sounding object of a powerful gaze. His resistance to that power predates it, indeed is the condition of possibility for a response to that power that is knowing, appositional, strategic. Nixon's interpellative call has damn near every institutional apparatus behind it: school, the seductive mystico-economic power of civil service in the form of "the force," and sanitation. And even if his insidious demand for recognition works in tandem with Diamond's multiply sourced feelings of guilt, the object resists here and in so doing simply rearticulates the condition of possibility of the liberatory. Nixon's attempt to reinitialize the "scene of subjection," to replicate the scene of his own subjection, is cut by another mode of organization, the (necessarily musical) theatre of objection, black performance as the resistance of the object.

This is to say, that black musical performance once again offers for us an instance of itself *as the ongoing reproduction of that which can be activated to disrupt* the reproduction of the conditions of capitalist production (the original [aim and] object of study for Althusser's famous essay even though now it is sometimes in danger of being reduced to another theory of the subject or of subjection). That which is invoked here, that which remains to be activated, is not merely some internalization of the outside as lost object but the always already given possibility of the exteriority of the inside, the becoming-object of the speaking, singing, commodified object. This becoming-object of the object,[4] the resistance of the object that is (black) performance, that is the ongoing reproduction of the black radical tradition, that is the black proletarianization of bourgeois form, the sound of the sentimental avantgarde's interpolative noncorrespondence to time and tune, is the activation of an exteriority that is out from the outside, cutting the inside/outside circuitry of mourning and melancholia. Here utopia is reconfigured in a morning song, at morning time by the sound of a moan of pain and joy. "Ghetto Supastar" carries that sound, a mo'nin' for morning, as the beginning of a day made even closer when the dead awaken to a kind of *working for*,[5] the working for all of the living, all who have lived, all who shall live.

Notes

1. In fact, the listening/questioning invoked here is that of my teacher Masao Miyoshi. He spoke of this at the 1999 MLA convention at a panel on utopian thinking organized by José Estéban Muños. The present paper is a revision of remarks I made at that conference.
2. In this paragraph the occasion that prompted this paper is alluded to again. At the 1999 MLA convention, then MLA President Edward Said organized several symposia on the theme "Scholarship and Commitment." My thinking here is bound up with the connection between Miyoshi's experience of music and my own, the connection between these and Said's desire for music, a desire he has written of and played for so long. Note, here, that I know that I should but can't apologize for all the ungainly traces you'll come across of the occasions (and their aurality) that prompt this essay. I mean I know I should shut my mouth but we're talking about a musical performance!
3. Here are the relevant passages from Douglass. Time and space permit only an indication of the connection between "heart-rending shrieks/horrid oaths" and that singing which "corresponds to neither time nor tune":

   > I have often been awakened at the dawn of day by the most heart-rending shrieks of an own aunt of mine, whom he used to tie up to a joist, and whip upon her naked back till she was literally covered with blood. No words, no tears, no prayers, from his gory victim, seemed to move his iron heart from its bloody purpose. The louder she screamed, the harder he whipped; and where the blood ran fastest, there he whipped longest. He would whip her to make her scream, and whip her to make her hush; and not until overcome by fatigue, would he cease to swing the blood-clotted cowskin. I remember the first time I ever witnessed this horrible exhibition. I was quite a child, but I well remember it. I never shall forget it whilst I remember anything. It was the first of a long series of such outrages, of which I was doomed to be a witness and a participant. It struck me with awful force. It was the blood-stained gate, the entrance to the hell of slavery, through which I was about to pass. It was a most terrible spectacle. I wish I could commit to paper the feelings with which I beheld it.
   >
   > [...] Aunt Hester had not only disobeyed his orders in going out, but had been found in company with Lloyd's Ned; which circumstance, I found, from what he said while whipping her, was the chief offence. Had he been a man of pure morals himself, he might have been thought interested in protecting the innocence of my aunt; but those who knew him will not suspect him of any such virtue. Before he commenced whipping Aunt Hester, he took her into the kitchen, and stripped her from neck to waist, leaving her neck, shoulders, and back, entirely naked. He then told her to cross her hands, calling her at the same time a d—d b—h. After crossing her hands, he tied them with strong rope, and led her to a stool under a large hook in the joist, put in for the purpose. He made her get upon the

stool, and tied her hands to the hook. She now stood fair for his infernal purpose. Her arms were stretched up at their full length, so that she stood upon the ends of her toes. He then said to her, "Now, you d—d b—h, I'll learn you how to disobey my orders!" and after rolling up his sleeves, he commenced to lay on the heavy cowskin, and soon the warm, red blood (amid heart-rending shrieks from her, and horrid oaths from him) came dripping to the floor. I was so terrified and horror-stricken at the sight, that I hid myself in a closet, and dared not venture out till long after the bloody transaction was over. I expected it would be my turn next. It was all new to me. I had never seen anything like it before [...].

The slaves selected to go to the Great House Farm, for the monthly allowance for themselves and their fellow slaves, were peculiarly enthusiastic. While on their way, they would make the dense old woods, for miles around, reverberate with their wild songs, revealing at once the highest joy and the deepest sadness. They would compose and sing as they went along, consulting neither time nor tune. The thought that came up, came out—if not in the word, in the sound;—and as frequently in the one as in the other. They would sometimes sing the most pathetic sentiment in the most rapturous tone, and the most rapturous sentiment in the most pathetic tone. Into all of their songs they would manage to weave something of the Great House Farm. Especially would they do this, when leaving home. They would sing most exultingly the following words:—

"I am going away to the Great House Farm!
Oh, yea! O, hea! O!"

This they would sing, as a chorus, to words which to many would seem unmeaning jargon, but which, nevertheless, were full of meaning to themselves. I have sometimes thought that the mere hearing of those songs would do more to impress some minds with the horrible character of slavery, than the reading of whole volumes of philosophy on the subject could do.

I did not, when a slave, understand the deep meaning of those rude and incoherent songs. I was myself within the circle; so that I neither saw nor heard as those without might see and hear. They told a tale of woe which was then altogether beyond my feeble comprehension; they were tones loud, long, and deep; they breathed the prayer and complaint of souls boiling over with the bitterest anguish. Every tone was a testimony against slavery, and a prayer to God for deliverance from chains. The hearing of those wild notes always depressed my spirit, and filled me with ineffable sadness. I have frequently found myself in tears while hearing them. The mere recurrence to those songs, even now, afflicts me; and while I am writing these lines, an expression of feeling has already found its way down my cheek. To those songs I trace my first glimmering conception of the dehumanizing character of slavery. I can never get rid of that conception. Those songs still follow me, to deepen my hatred of slavery, and quicken my sympathy for my brethren in bonds. If any one wishes to be impressed with the soul-killing effects of slavery, let him go to Colonel Lloyd's plantation, and, on allowance-day, place himself in the deep pine woods, and there let him in silence analyze the sounds that shall pass through the chambers of his soul,—and if he is not thus impressed, it will only be because "there is no flesh in his obdurate heart." ([1845] 1986:51–52, 57–58)

4. I must indicate, however briefly, that I am indebted to but deviate from Jacques Derrida's notion of "the becoming-objective of the object" (see Derrida 1997:31).
5. This is that working for to which Miyoshi is devoted. This essay is dedicated to him if he'll have it.

References

Adorno, Theodor W.

1982 "On The Fetish-Character in Music and the Regression of Listening." Translated by Maurice Goldbloom. In *The Essential Frankfurt School Reader*, edited by Andrew Arato and Eike Gebhardt, 270–99. New York: Continuum.

Althusser, Louis

1971 “Cremonini, Painter of the Abstract.” In *Lenin and Philosophy*. Translated by Ben Brewster, 229–42. New York: Monthly Review Press.

1971a. “Ideology and Ideological State Apparatuses (Notes towards an Investigation).” *Lenin and Philosophy,* translated by Ben Brewster, 127–86. New York: Monthly Review Press.

Butler, Judith

1997 *The Psychic Life of Power: Theories of Subjection*. Stanford: Stanford University Press.

Derrida, Jacques

1997 *Resistances of Psychoanalysis*. Translated by Peggy Kamuf, Pascale-Anne Brault, and Michael Naas. Stanford: Stanford University Press.

Douglass, Frederick

1986[1845] *Narrative of the Life of Frederick Douglass, An American Slave*. London: Penguin Books.

James, C. L. R.

1977 *The Future in the Present*. London: Allison and Busby.

Mackey, Nathaniel

1986 *Bedouin Hornbook,* Callaloo Fiction Series, Volume 2. Lexington: The University Press of Kentucky.

Marx, Karl

1975[1844] “Economic and Political Manuscripts.” In *Early Writngs,* edited by Quintin Hoare, translated by Gregor Benton, 279–400. New York: Vintage Books.

1990[1867] *Capital,* Volume I. Translated by Ben Fowkes. New York: Penguin Books.

Michel, Prakazrel “Pras,” with kris ex

1999 *Ghetto Supastar*. New York: Pocket Books.

Negri, Antonio

1991 *Marx Beyond Marx: Lessons on the* Grundrisse, Edited by John Fleming, translated by Harry Cleaver, Michael Ryan, and Maurizio Viano. New York: Autonomedia/Pluto.

Pras

1988 “Ghetto Supastar.” In *Ghetto Supastar*. New York: Ruffhouse/Columbia Records CK 69516.

Robinson, Cedric J.

2000[1983] *Black Marxism: The Making of the Black Radical Tradition*. Chapel Hill: The University of North Carolina Press.

A Tribe Called Quest

1991 “Jazz (We’ve Got).” In *The Low End Theory*. New York: Jive Records 1418-4-J.

Mendicant Erotics [Sydney]

a performance for radio

Ellen Zweig

Some Notes on the Production

When Andrew McLennan of ABC Radio Australia asked me to create a piece for his *Listening Room* series,[1] I immediately thought of using the opportunity to begin work on the performative aspect of a novel I'd been planning. This novel, *Mendicant Erotics*, is about a woman who becomes a beggar after studying the *Esanashamiti*, a book of rules for Jain wandering monks. Traveling from city to city, she follows these rules, which specify six geometric patterns through which the mendicant can approach the houses in a village for begging. Disoriented by sensation, she finds a prosthetic architecture, buildings that take over the functions of her body or through which she can project her secret and intense desires. A wandering mendicant, begging for erotic experience, her desires are met in the arms of the built environment. Hers is a geometry of the trajectories of traveling, memory, and architecture. Fantasy narratives evoked by city spaces are the alms she extracts from each place. These narratives form the novel, a sort of contemporary picaresque adventure which is also a philosophical treatise on the relationships among bodies, buildings, city spaces, and spiritual endeavors.

I wanted to experiment with the method that would later give me material for the novel and also to make a piece specifically for Sydney—a site-specific radio piece. Since I could not wander the streets of Sydney in advance of my visit there, I developed a collaborative system for getting the raw material. Thus, the performative aspects of the piece reached across miles of land and ocean, traveling before me, as I would travel to the land down under. I wanted, first of all, an Australian voice, male, very much from that place, so I asked Andrew McLennan (a radio personality well-known to the Australian listening public) to help. I drew a square on the map of Sydney and faxed it to him with the following instructions:

1. Record about five minutes of ambient sound at each site.
2. Describe the site, especially any sensory information, details of sight, sound, smell, touch, taste.

3. Describe especially the quality of light. Also, describe the architecture at the site and any architecture you can see from the site. What do the buildings look like and how do they interact with the light?
4. If you can enter a building, go inside. Record ambient sound inside. What is the difference between inside and outside?

Andrew trekked through the pouring rain of winter in Sydney, always astonished at a break in the clouds. The resulting tapes were just what I needed to create the male character in the dialog. He is very Australian, knows Sydney well, sometimes sounds like a tour guide; is wildly enthusiastic, often like a small boy who's seen something lovely or tempting; sometimes rambles on, inarticulate, stumbling, but always eager to please. His language is, for the most part, demotic. Sydney is his place and he feels very much at home commenting on its sensations. In contrast, the female character seems to speak in a void; sometimes she's so caught up in her fantasies that she seems in another place; sometimes she's right there with him, but critical or dejected. She's an American voice and not at home in Sydney. Her language is mostly hieratic and poetic. She's unmoored.

The ambient sounds in the piece are from the four sites where the male and female characters meet. The interludes are little transitional sound collages that indicate movement from one site to the next. The piece begins and ends with the female voice, solo, in the nowhere void of traveling. (Right now, she's drawing on the map of Hong Kong.)

Mendicant Erotics

[Sydney]

The Map
I was swimming in a liquor of black night. I was up and walking around and my body felt distant, almost alien. It was late at night, very late and I couldn't sleep. Again. And just like the other times, there was a high gloss, almost black sheen to my thoughts. How can I describe this isolation—so intense that if someone had entered my room, it would surely have seemed as if I was no longer there? I felt light, almost floating, but heavy with doubt. My mind was on fire and my body was tingling, aching as though pulled by unknown forces.

Wandering around the kitchen, I opened the refrigerator door *(sounds of water and fog horns begin here and increase in volume and intensity throughout this introductory text),* found a shriveled potato without aluminum foil and some peas in a plastic bag. I sat down on the floor with the peas, opening each pod and looking closely at the rows of round green peas inside. Some were perfectly round and others were distorted, squeezed and shaped, almost square. I ate one, sweet and tangy and raw. Ate another—tasteless and mealy. Swept everything back into the bag, pods and all, and put the bag back into the fridge. The cool wet air lingered on my face and chest. But I didn't feel less apt to evaporate, losing my body into some dream of another place...

I live in several layers of phantom architecture. Whenever I feel ecstasy about my condition, I'm reminded that I don't know my way around even the dim hallways of my apartment. Although I can manage to predict the

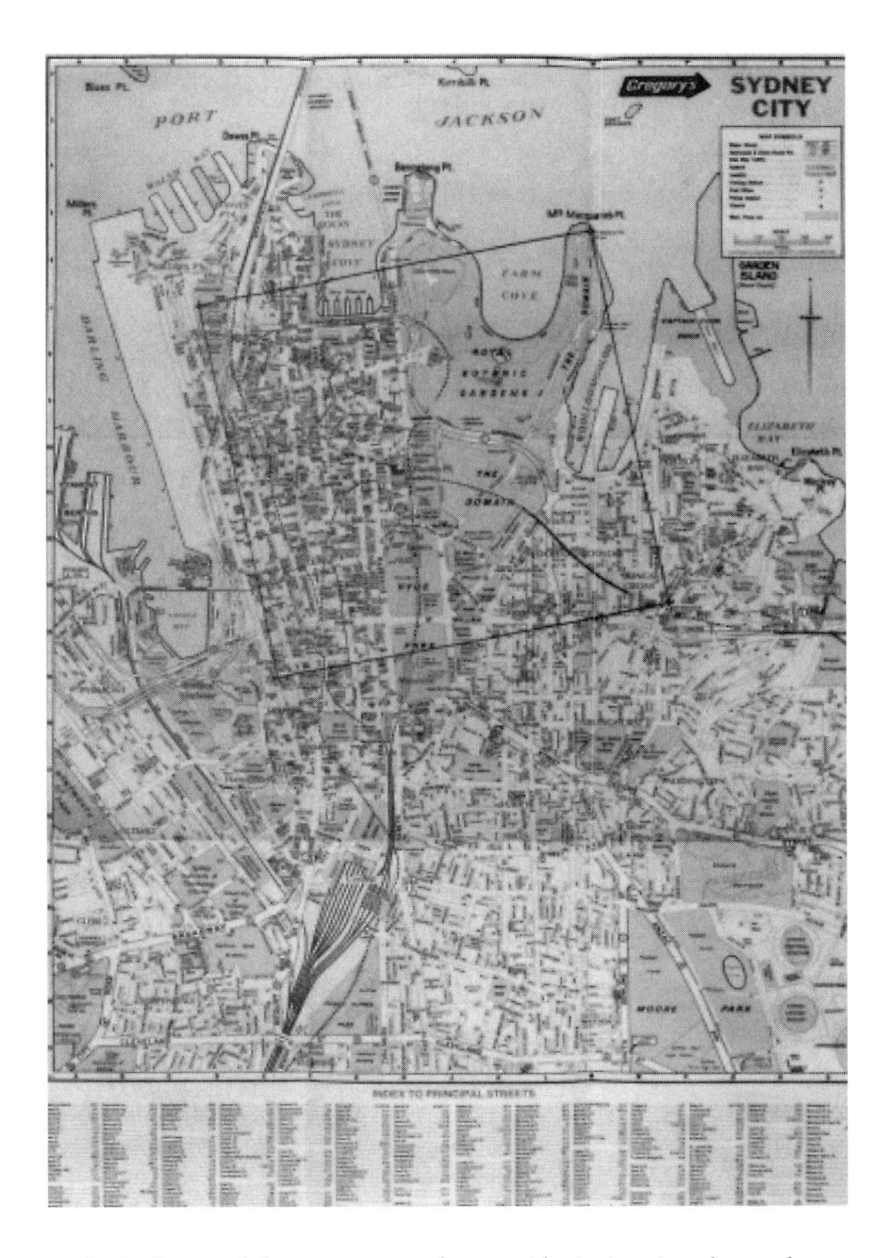

cracks in the wood floor soon enough to avoid tripping, I can't ever know just when that uncanny moment will arrive.

Surrounded by useless Victorian optical instruments—magic lanterns, postcard projectors, phenakistoscopes, and zoetropes—I am helpless at 4:00 AM when my house becomes haunted. I feel unmoored, without bodily substance; every room's a stage for certain unnatural communications with the dead.

And as if this experience were still not intense enough, my skin begins to crawl, the hair stands up on the back of my neck, and every now and then, my right foot aches just at the point where I fractured it several years ago. To disorient and dematerialize the world, to create a space in which architecture becomes prosthetic, I begin to imagine a nomadic life. Perhaps the intensity of my isolation would be tempered outside in the forbidden space beyond the confines of my well-defined world.

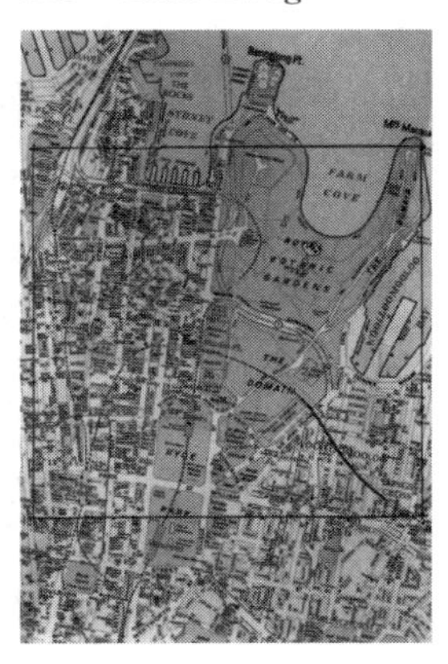

Surely there is a building whose dimensions have been made unknowable by the hallucinatory power of travel. There's nothing incongruous about it, nothing strange about the way I suddenly know what to do. I pack my bag, everything, even the flame-colored silk scarf that I never wear.

I want a life requiring no maintenance. Listen to the sounds of the night. Garbage trucks clanking and revving their engines at 2:00 A.M. A woman screams and somebody laughs. Wind through a tunnel of buildings. My ears automatically record what my mind ignores. The wood floor looks like a sand-colored carpet; the traffic sounds like the ocean. I'm in so many places at once and nowhere at all.

I'm still at home, alone, late at night. The light leaks in over the windowsill with full details of the outside world. It's a camera obscura in here, and the image is upside down. The light tracks the wall's height—goes abruptly blank—and I turn away, wandering from room to room with nothing to do and nothing to think about. My mind's a blank. Camera rasa.

Then there's the moment of departure from a strange place, which stops me right in my tracks. I can see it so vividly now—the spot where people go swimming or go boating for pleasure. I'm ready for the structures of astonishment—take me over, I say—outloud.

I'm skimming on the surface of the world, while the smell of jasmine and wild indigo go by so quickly. I try to catch and keep that smell with my controlled silent contortions. Even in the watercolors of the city, there's no map for getting lost.

This trip will be dominated by three phrases: I hold out my hand for them. All through the day they slide by.

I am the one who is begging at the door of the erotic. Will you touch me, untouchable as I have become? I am the one who holds the whole world in my outstretched hand. What will you give me in return? I leave New York with nothing but my night-wandering, empty mind. I have no idea what I will find.

I follow the paths of the wandering monks in the Jain book of rules, the *Esanashamiti*. I must walk the streets of the city in one of six geometric patterns. Here on a map of Sydney, I draw a square.

interlude: *sounds: clapping; woman's laugh; footsteps; male voice: "that would have been a nice sound"; clapping.*

Site 1: Mrs. MacQuarie's Chair
(sounds of rain, tourists, birds)

MALE: This is the twenty-first of June, that's to say it's the, it's the shortest day of the year.

FEMALE: Several layers of phantom architecture wobble and coalesce in the dank atmosphere of the early morning. There is an ecstasy about the pale green light.

MALE: In fact, I'm sitting on it. It's, it's just one of those features of the Sydney landscape that naturally forms itself into chairs. Old sandstone rocks here have been worn away over the years and create little caves.

FEMALE: I stand just inside the arched doorway, a stage for certain unnatural longings. A woman in a brown fur coat brushes by my hand but her touch is still not intense enough to obliterate the feeling of open space.

MALE: But for the moment we're here.

FEMALE: The dark cool window across the lobby is surely meant to disorient and dematerialize the passage through the door. Summoning a profound resistance, I turn left, entering a forbidden space beyond the elevator shaft.

MALE: The Moreton Bay Fig trees and the Port Jackson Figs are around here. Huge buttresses and wonderful sort of spreading limbs that sort of reach out for miles.

FEMALE: I was swimming in a liquor of black night, skimming on the surface of the world.

MALE: Curiously enough, I mean, the place is full of mad contrasts.

FEMALE: This building has a high gloss, almost black sheen.

MALE: Occasionally you can hear them sort of shouting their commentary *(laughs)* over megaphones to the astonished crowd.

FEMALE: Your imagination is still not intense enough to evoke a world without architecture.

MALE: These sort of sandstone buttresses that come down into the water. And right now the sun is just coming out again in which it does its most fantastic thing where it's catching, sort of in that silver way, the surface of the water. Water's very calm actually, at the moment. Sort of rippling along, there's no, there's no...it's a calm harbor.

FEMALE: Am I ten thousand years too early or only six thousand miles away?

MALE: Looking back around the point though... Farm Cove is also sort of...it's a walled-in cove. Just about all the coves here in Sydney now are walled-in. They've got nice solid sandstone walls that the water butts up against...it's...it usually creates beautiful patterns. We're not listening to those at the moment, but it's typical of this harbor that you can hear beautiful water sounds—as the water, um, reaches high tide, it comes up and flows and eddies through the rocks.

FEMALE: I have the sad conviction that I will always be haunted by architecture.

MALE: Like the kind of back of some huge prehistoric monster kind of emerging out of the ground.

FEMALE: I *could* take you to the body of water whose dimensions have been made unknowable.

MALE: It's a sort of curious, rich, moisty kind of smell.

FEMALE: There's nothing incongruous about it, nothing strange about swimming, you know.

MALE: Rain's coming in a bit more strongly now. Here you can hear it sort of hitting the rock and the leaves of the Moreton Bay Fig. This is, this is where, you know, Lady MacQuarie and the early colonial people would've loved to have come for picnics.

FEMALE: In the bright days when we were unacquainted with the world.

interlude: *sounds: high-pitched squeal; male beggar's voice: "can you spare change"; footsteps; someone shouts "hey"; male voice: "one of those insidious sounds, isn't it?"; motorcycle; woman's laugh.*

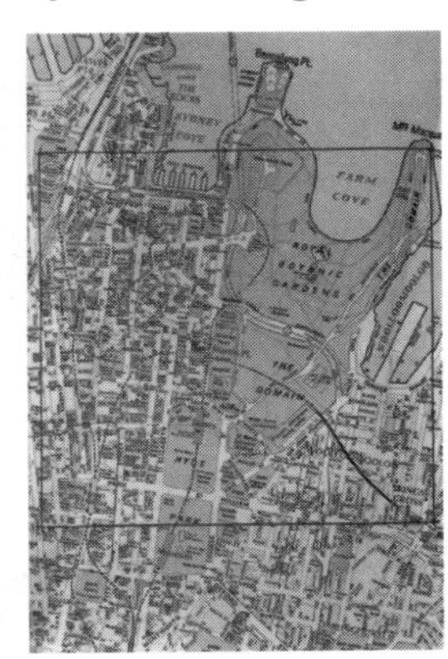

Site 2: King's Cross
(traffic, shouting, music)

MALE: Yeah, well, it's not without some trepidation that I stand here on the corner of Darlinghurst Road, Bayswater Road, Williams Street, Victoria Street, in fact, the region that is known as King's Cross.

FEMALE: To touch my own hand here, there's nothing incongruous about it, nothing strange.

MALE: You can hear music blazing forth from the cafés, the strip clubs, the joints, and the motor cars. There's just music just pumping everywhere.

FEMALE: Suddenly, I'm sitting in a pink satin parlor enmeshed in the structures of astonishment.

MALE: I'd hesitate to go into any of these buildings here.

FEMALE: I smell tuberose and jasmine and wild indigo.

MALE: As I said I don't fancy going into any of the buildings. I don't particularly want to see a strip joint; I don't want to go into Condom Kingdom; I don't fancy any of the fast food; I just had a coffee down on the El Alamein fountain which is one of the nicer parts of the drag. What I do fancy doing though is, if I can find the time, is to go inside the Hotel Capitol, formerly the Crest Hotel; the Hotel Capitol has now been taken over by a Korean consortium and I hear that inside there now on the first floor you can go in there and have a full Korean-cum-Japanese bath. There's a bathhouse there with full Japanese sauna, massage, all kinds of interesting treatments and that's what would attract me I think, that's what I'd like, want to know what was going on here.

FEMALE: Even in the watercolors, the ruins of this room look too dusty.

MALE: *(door slams, water sounds, locker doors clacking)* Yeah, well, I can't actually see anything because I went inside the sauna and I had to go in there without my glasses *(laughs)*, otherwise they'd steam up.

(water sounds continue)

FEMALE: I was swimming in a liquor of black night, skimming on the surface of the world.

MALE: The whole place is tile, and it's slippery. I feel the...I'm naked, of course, I had to take my clothes off and it's slippery. There are tiles. There's a big sort of mural on the wall depicting kind of waterfalls, and various other sort of watery retreats. Then, there are extra cubicles for sitting down and doing a sort of very special kind of wash. You can wash every part of yourself, have a shave, just get stuck into all the bits and pieces, all the pores, the crevices, the nooks and crannies on the body—

FEMALE: —lurching, unsteady, unmoored.

MALE: —and generally the sound is the sound of kind of hiss, the hissing of water, the gurgling of taps and sounds like that but just a permanent hiss of water flowing,

FEMALE: —tempted by demons, false memories and fantastic ghosts.

MALE: Yes, taps, taps and tiles; the sound of water running, various voices—*men's* voices.

FEMALE: This building has a high gloss, almost black sheen. It sticks in my throat.

MALE: Very much conscious of your skin. Skin in water...

FEMALE: —lurching, unsteady, unmoored.

MALE: This is the ideal position to observe the rest of King's Cross, because it's behind glass but looking down—onto the main *drag.* I hadn't expected this...

FEMALE: —tempted by demons, false memories, and fantastic ghosts.

MALE: Neon lights, live shows, girls, movies...that's Playgirls International, the X club, and further down there you can see magazines, videos, toys and novelties, souvenir discounts, the Kodak Express...

FEMALE: I was swimming in a liquor of black night, skimming on the surface of the world.

MALE: There's the Love Machine with live shows: girls girls girls and XXX movies, all lit up there in blue and red, flashing away garishly at you. Yeah, the perfect vantage point.

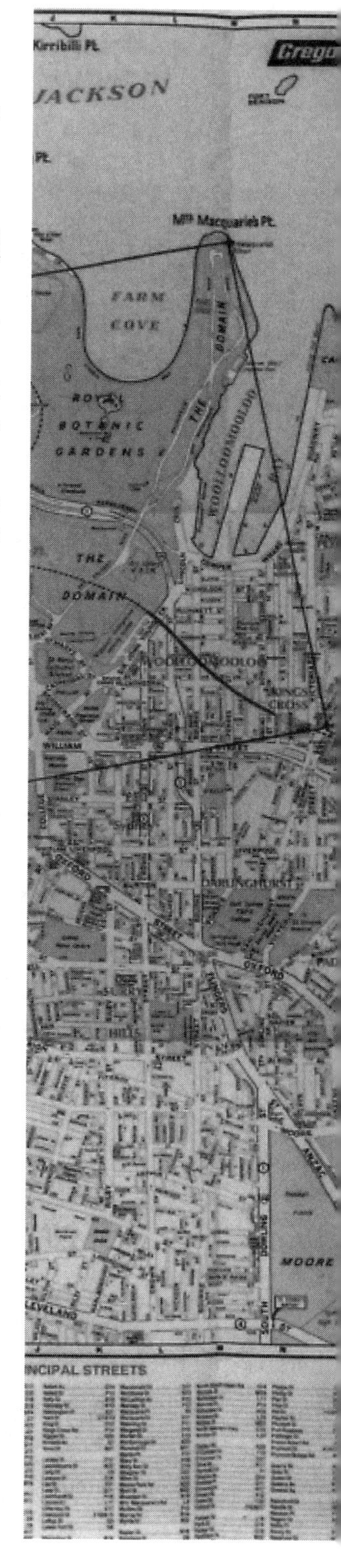

interlude: *sounds: clapping; squeal; woman's laugh; male voice: "that would have been a nice sound"; footsteps.*

Site 3: Sussex and Bathhurst Streets
(traffic)

MALE: We're here pretty well on the corner of Sussex and Bathhurst Street.

FEMALE: All through the day they slide by—memories and hallucinations.

MALE: Didn't even know it was here. It's extraordinary—it's almost like a temple.

FEMALE: You take me through the dark hallway and we enter several layers of phantom architecture to get to the motorcycle repair shop. I feel a sudden and unexplained ecstasy about the roar of engines embedded in so much metal.

MALE: So, everything around here is sort of higgledy-piggledy. There are glass towers sort of looming up, there're sort of...there's uh, ah well of course...

FEMALE: Each man holds a set of useless Victorian tools. I can see this is a stage for certain unnatural acts and I'm tempted to perform. But the light is still not intense enough for me to tell the difference between inside and outside. The men move their wrenches to disorient and dematerialize this room, which wobbles between a courtyard and a garage. Give me a forbidden space beyond the body's ache.

MALE: I'm always fascinated by this drainage system, you know, where you actually don't have down-drains to the ground. You simply have a hole in the drainage on the roof and a chain comes down and somehow or other the chain is meant to lead the water down into the ground. I...something I've never, never understood.

FEMALE: The men surround me in a pattern whose dimensions have been made unknowable. There's nothing incongruous about it, nothing strange about the effort it takes to pass through the maze of their arms and legs and to see again a bold shaft of light echoing down the side of a glass facade. It is as though the whole building is bathed in flame-colored silk.

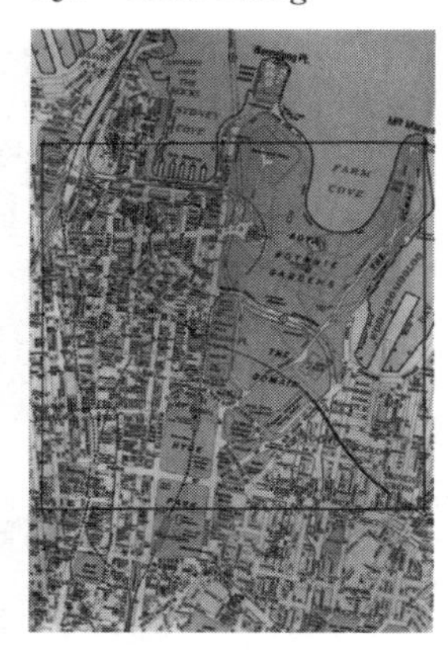

MALE: It functioned as the city night refuge from 1868 to 1972, so I feel that we're really on the spot.

FEMALE: Give me the courage to recognize the moment of departure from a strange place.

MALE: It's not a house for beggars anymore.

FEMALE: The taste of metal on my tongue comes from the smell of motorcycle oil and the structures of astonishment left over from my *(high-pitched squeal)* close call.

MALE: But the thought of it being a place for the homeless and the poor, those without food, is interesting.

FEMALE: Jasmine and wild indigo twine around the lamps of this room.

MALE: Yeah, we've just walked into the veranda of the Flying Angel Seafarer's Center and it's in the precincts of this old school built in the late 1800s. You can just go inside. Oh, it's a beautiful tesselated tile and it says here "welcome all seafarers," which I feel is only appropriate for...certainly for me coming from a great seafaring family. And here in the anteroom here, we see...ah it's great, look...

FEMALE: The walls are closing in, in controlled silent contortions. I'm tired, so tired, I need to sit down.

(interior sounds: talking and music)

MALE: *(a more intimate tone of voice, quiet)* I just walked into a beam of sunlight, which is always a pleasant experience in a dark building, and I thought...I was a little bit disoriented by it. I look outside and I actually find it's reflected sunlight. The sun is actually being reflected off the building to the south here and it's coming back in directly through one of these windows. Windows are all barred. Whether that was to protect the windows themselves? to protect the children? or to keep them in, perhaps, so that they couldn't escape.

interlude: *sounds: footsteps*

Site 4: Observatory Hill

(rain, birds)

FEMALE: I have the sad conviction that this will never end and the world will grow stronger in its intricacy.

MALE: Here standing on the top of Observatory Hill now. There's another break in the clouds *(laughs)* for some inexplicable reason and we've put ourselves in the rotunda just in front of the Sydney Observatory. It's just behind us here. But, this view is almost impossible to describe, because there's just so much in it. But what immediately takes the eye is the different perspective of the Harbor Bridge and it's being raked with light at the moment from the break in the clouds. Because it's wet, it's glistening, and it's been glistening. The difference in light and shade on the struts and on the supporting columns of the bridge are sort of really dramatically etched there into the skyline, and it's from this hill that the bridge actually launches itself. You can hear there...there's this big silver train—it's glistening in the sunlight too as it's just making its way over the bridge now. It emerges from the tunnel just down below here, just on the edge of this hill.

FEMALE: I'd do anything I could to conceal the fact of it, of the minutiae and everything that brushes up against me.

MALE: Huge towers with great curving lines and massive ferris wheels leaping up out of and perched on top of curious supports and the mad entrance to it all, the sort of Art Deco world of...with a big face inside it, you know, leering at everyone that goes in.

FEMALE: Architecture constantly creates structures of astonishment, doors going in and out, windows for views, long hallways with sudden narrow turnings...

MALE: —but here once again the clouds are being raked with light.

FEMALE: The light leaks in over the windowsill with full details of the outside world.

MALE: But here it's nice, it's verdant and green on this hill. It's a park. As I say behind us, the early Sydney Observatory—we can't get into it at the moment, but with a bit of luck...I've made an appointment to get in and we'll see what happens later when we have a shot at that. But we're standing here on butcher's grass, false grass in the rotunda. I've no idea why they put grass here, but it stinks of urine. So it's a, presumably it's used as a public toilet by the homeless of Sydney.

FEMALE: The path is choked with jasmine and wild indigo.

MALE: Oh, the light's just coming out again now and things are starting to glisten again—all the leaves in the Moreton Bay fig trees are just glistening with the rain on them and the grass is glistening too—everything is sort of glistening which has that marvelous fresh feeling about it. Wind's dropped down a little too. That's nice. The support structures for the bridge are sort of standing out in their kind of Aztec Deco form again. People have likened the Harbor Bridge to a kind of huge Aztec sort of rising sun. It does have a tendency to look a bit like that sometimes.

FEMALE: The light tracks the wall's height—goes abruptly blank—and I turn away...

MALE: Ah, it's almost nice being here now.

FEMALE: You swim with controlled silent contortions, licking the water from your lips with every turn of your head.

MALE: And the two copper domes that—like two lovely breasts sort of sitting here on top of the...on top of this hill.

FEMALE: When I was a child, I was told to embroider a small house that looked just like the house we lived in then.

MALE: And with the tower behind it, in fact it's a...oh, well, yes it's, it's one of those ball towers. So, it would have been...it would have signaled the time to the ships. It would have been...they would have dropped that ball down at one o'clock, yes, one o'clock. They would have dropped the ball down at one o'clock for all the ships to set their chronometers to.

FEMALE: If the embroidered house had only belonged to a larger pattern...

MALE: Terrific inside those domes—that might be interesting—and see all the old sort of optical things and the telescopes are still there.

FEMALE: Even in the watercolors, the house is isolated, framed as a single house in a desolate landscape.

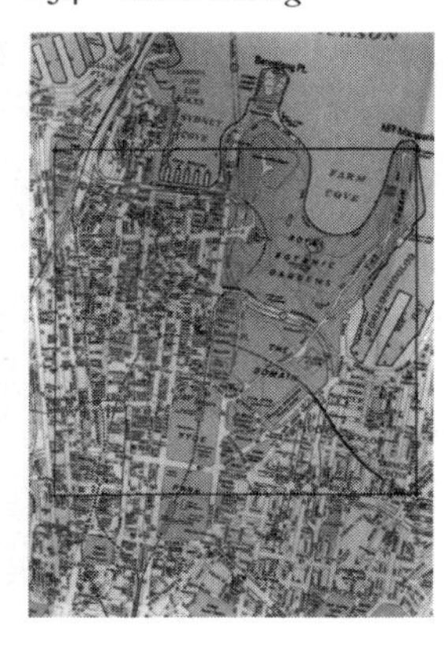

MALE: *(footsteps)* It's actually a marvelous sensation swinging such a huge mass around and it seems to be skipping over the surface of the dome.

(echoing clanking, low hum—this is the sound of the dome turning on ball bearings—it continues...)

FEMALE: So many possible houses—all through the day they slide by.

MALE: *(conspiratorial and intimate tone of voice)* In terms of, in terms of erotics I suppose, needs to be mentioned that in both of these domes, there are wonderful slits, slits that open up to the sky. It's sort of as if one's inside something, isn't it? Inside some body. Mmm, some great organism, these big slits, vulva perhaps. It might open up and disgorge us or maybe swallow us—one's not sure.

interlude: *sounds: squeal and clanging; woman's laugh; male voice: "that would have been a nice sound"; squeal and clanging.*

coda

Dominated by three days of torrential rains, the ocean beckons. At least 28 men on motorcycles—all through the day they slide by—accompany me to an entrance, an exit, another journey over water, in the bright days when we were unacquainted with the world.

Note

1. *Mendicant Erotics* was created August 1995, aired on Radio Australia on 23 October 1995, and aired at the Adelaide Festival of the Arts March 1996.

Casual Workers, Hallucinations, and Appropriate Ghosts

Toni Dove

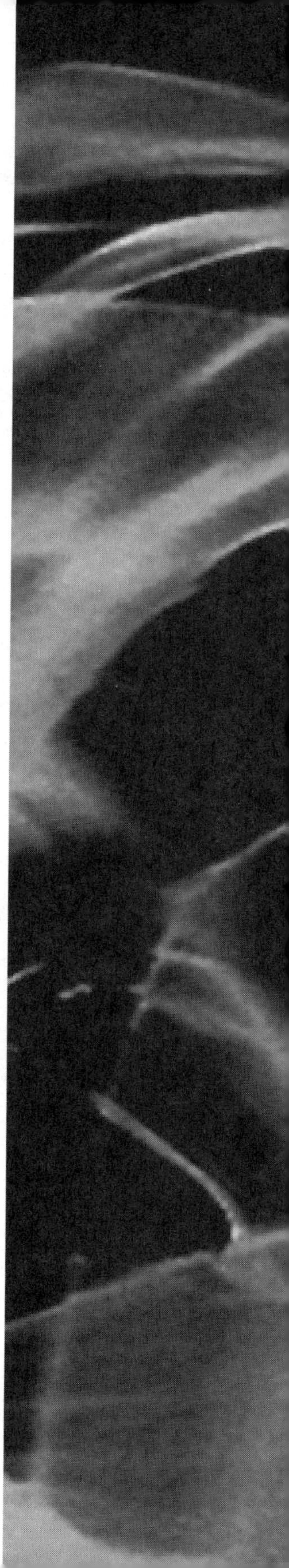

Casual Workers, Hallucinations, and Appropriate Ghosts began as a radio soundtrack. It was an expansion of the installation *Mesmer — Secrets of the Human Frame*, a part of the 1990 *Art in the Anchorage* exhibition sponsored by Creative Time, Inc. The radio version elaborated on the theme of the construction of identity over the course of a century as seen through cultural representations of robots, androids, and cyborgs. The historical spectrum stretches from the industrial revolution to the technological revolution and attempts to examine these shifts in paradigm and to make sense of the ground that is currently moving under our feet. *Mesmer* had a number of incarnations: an installation, a radio play (New American Radio, 1991), an artist's book (Granary Books, 1993), and an essay in *TDR* (1992).

The piece which follows, the caboose of the radio play, went on to become the soundtrack for a 1994 video installation completed for the lobby of a movie theatre in Times Square. The installation—commissioned by Creative Time, and the 42nd St. Development Corporation for the 42nd Street Art Project—presents a metamorphosis from the theatre of hysteria (represented by Charcot's 19th-century photographs of hysterics reflected in bubbles floating around the head of a dancer acting out his choreography of hysteria) to the choreography of the female heroines of contemporary martial arts films. It is accompanied by a narrative of disturbances in the fabric of human intimacy followed by a three-minute symphony constructed entirely of screams. The story implies a female subjectivity that reclaims certain familiar narratives from popular culture. The piece was situated at the end of a series of adult video stores. Video and sound were seen and heard from the street.

These projects, a fugue of hybrid emergence, define a method of working in which complex material is given multiple lives. Each piece has many incarnations in different media and modes of address that allow the reader/viewer/listener, as well as myself as author/producer/director, to assimilate material that does not lend itself gracefully to the brevity of a sound bite. As every story is both a door closing and the opening of new fingers of thought, there is a symmetry to presenting this final state of an organic sequence in the publication where its first version, itself incorporating other states, was presented.

1. Section of the background image from the poster for Casual Workers, Hallucinations, and Appropriate Ghosts *by Toni Dove. Part of the soundtrack for the 1994 video installation is featured on the* TDR *compact disk,* Voice Tears. *(Image by Toni Dove)*

2. Casual Workers, Hallucinations, and Appropriate Ghosts, *a video installation by Toni Dove, was sponsored by Creative Time, Inc., and the 42nd St. Development Corporation, and appeared at the end of a series of adult video stores. (Photo courtesy of Toni Dove)*

Radio Play Text by Toni Dove and Judy Nylon
Performed by Judy Nylon
Sound Design by Toni Dove and Dana McCurdy
Created at Harvestworks Digital Media Arts, New York, NY

The story refers to two short novellas by Yasunari Kawabuta published as *House of the Sleeping Beauties* (1969), as well as various contemporary science fiction stories.

The piece begins with a solitary voice telling a story. The narrator's voice is female—husky and erotic in quality. Approximately one-third of the way through the text, a bed of sounds fades in slowly. The sound of crickets, bugs—sounds of nature. This is

3. Installed in the lobby of a movie theatre in Times Square, Toni Dove's 1994 Casual Workers, Hallucinations, and Appropriate Ghosts *presents a "metamorphosis from the theatre of hysteria [. . .] to the choreography of the female heroines of contemporary martial arts films." (Photo by T. Charles Erickson)*

punctuated by occasional shrieks that sound at first like large birds and gradually like human screams. The mental image is of a huge wire birdcage populated by creatures which may be birds or may be human or part human. Gradually a repetitive abstract sound begins to build like an anxiety attack—a sound not unlike rapid breathing, but heavily processed electronically. As the text moves into paragraph two and begins to refer to science fiction stories, the sound environment becomes more rhythmic, synthetic, and mechanical. There is an ominous tension. It builds until the final statement of the text is spoken in the clear.

I remember reading this story which starts with two old men who have been very successful, powerful, now retired—impotent. They meet in a gentlemen's club every afternoon like two declawed tigers. Each is the other's witness when they talk about their memories. One man tells of a house of prostitution he's been visiting. It's by the sea—beautiful, understated. The girls are all very young. They're drugged—totally unconscious—like empty bodies or blank canvases he can project his memories on. He can relive great affairs, great passions. The experience jars memory, the smell of very young flesh—to poke it and have it come back springy and plump with that slightly milky smell that young humans have. He's completely hooked on this. He begins to visit the same girl again and again. But—after he's poked her a little bit and smelled every area of her body and explored this and that—well—there's no relationship there. He realizes that he can reach through the haze of the drug by inflicting subtle pain—not beating her, but something that will leave no marks—a small thing, like a slight pressure along the meridians of her body. I suppose that when the girl feels pressure—because he's squeezed or

4. An image from Toni Dove's 1994 video installation, Casual Workers, Hallucinations, and Appropriate Ghosts, *in Times Square. "Except this woman is having a pain-stimulated relationship with someone she never sees or actually knows. She's bound to have some kind of bleed-through in her everyday life." (Photo courtesy of Toni Dove)*

hurt that knuckle or joint before—she's going to move. He creates a relationship with her through subtle pain. The girl wakes up in the morning with no memory of this. She's been the dutiful daughter. Except this woman is having a pain-stimulated relationship with someone she never sees or actually knows. She's bound to have some kind of bleed-through in her everyday life. If you flinch every time someone reaches for your neck, at some point you're going to ask yourself why.

It reminds me of stories I've read—science fiction, or speculative fiction, let's call it, because it's so close to true—about pleasure models, meat puppets they call them. They have neural bypasses which allow them to function as prostitutes with no memory of the experience. This one woman is working as a meat puppet to earn money to have her nervous system jacked up to become a street samurai, a kind of hired gun with bionic reflexes, and she starts to have bleed-throughs of brothel experiences that come through like the faint stains of a bad dream. It's a classic female catch-22: to have the tools she needed to make her professionally efficient, she had to have cash. So, she works as an unconscious pleasure model. It was like blank time—a matter of coming to and finding you're a little bit sore and wondering, Where the hell have I been?

There is a wailing, spiraling scream that starts from a distance and snakes into close-up; it begins as a processed sound and ends as a scream—raw and recognizable. It is the prelude to a three-minute choreographed "symphony" of eight tracks of multiple screams—some processed, some more natural. Each track has a pattern or rhythm that builds towards a climax. The finale is a crescendo of shrieks and screams that spiral in pitch-shifting delays, leaving a pure vocal note of operatic and melodic character floating away like smoke in the wake.

References

Dove, Toni

1992 "Mesmer: Secrets of the Human Frame." *TDR* 36, 2 (T134):62–76.

1993 *Mesmer, Secrets of the Human Frame.* New York: Granary Books. Limited edition: 50.

Kawabuta, Yasunari

1969 *House of the Sleeping Beauties.* Tokyo: Kodansha International Limited.

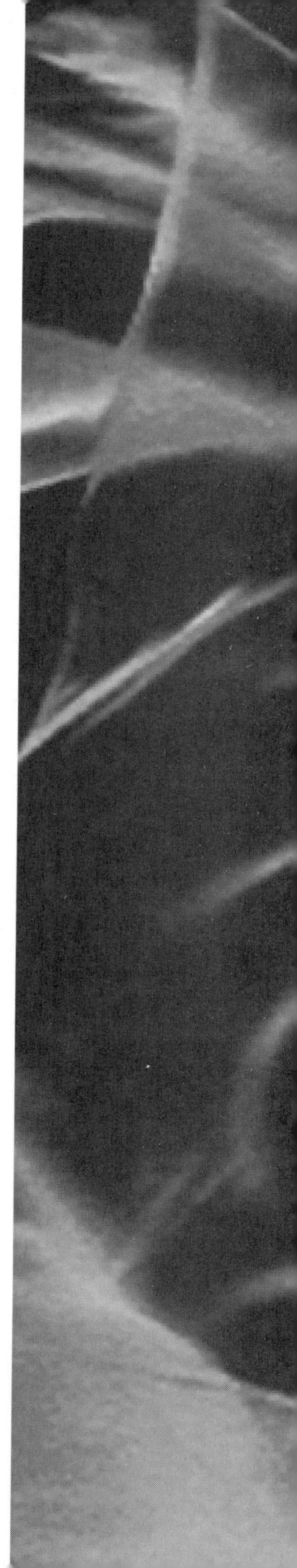

Hotel Radio

excerpt from the play by Richard Foreman

What I can bring back from my day exploring the city— ?
It vanishes.
Therefore, the city might have been endless. On the other hand, it might have been a disappointment.
That is one of the reasons I so miss having a radio in my room.
If there was a radio in my room I might, now, turning it on for myself, hear—intuition-wise—what I missed, or lost, in my meticulous exploration turned back toward me.
Nothing like memory you understand, but instead
like a broken self, a broken me,
and in those cracks
the wind of real things at last, through a radio—
in here—unheard.
And so I re-imagine a world entire,
living the rest of my life forgetful,
asleep,
sensing whatever purpose I picked up when I slid backwards into the wrong door titled "obligation through this door" and I was in the lost and found department again
—but it could never decide whether it was the land of the lost or the truly—FOUND at last!

(In a certain hotel,
a certain radio was absent.
Once upon a time each room contained a radio.
Now, no radio in no room.
In a certain hotel
plans were made
to broadcast from a tower on the roof of the hotel—
radio broadcasts.
But such plans never came to fruition.
Nevertheless, the name of this hotel
was the Radio Hotel.
How often has a name been less than appropriate?
In this case one could understand why the name was chosen,

even though it was no longer appropriate.
Having once been to a certain extent an appropriate name
there was now a pause
in thought.
Through that pause,
thoughts from another space
bled—leaving the residue of a name.
Radio Hotel!
No blood staining yet the walls of the Radio Hotel.
Silence—radio reigned!
And the Radio Hotel
closed inside itself
the lost proclivity it broadcast
toward street wards
those who passed or entered
or passed through as guests in the
Radio Hotel.)

"Hotel Radio—hello, radio Hotel."

Can I help?

No. But help.

It's one of my favorite words. No. It's my favorite word.

I help whenever I get help.

The more times you can use the word help in a sentence, the more it helps.

Help myself.

Help yourself.

Help yourself to the word help, which is how I help myself.

(Pause)

Help yourself to some fruit.

I don't think I should eat right now.

(Pause)

One of the most potent ideas I ever had, ever, was the idea that in the center of the fruit was a pit, and the pit was the radio in the center of the fruit. And the whole fruit helps—the radio in the center of the fruit.

My ear: helps.

My ear was help also.

Does this help? My ear helps.

(Pause)

Have some fruit now.

Eat it, or let it turn into the radio that it is.

In the Hotel Radio, the fruit placed in bowls which sit on small tables in each room—no radios in rooms, but fruit in rooms, and in the center of the fruit, is a radio.

(Pause)

I imagine walking down the street and seeing the letters painted on the stone wall of the building I pass to spell the words "Hotel Radio." Then I imagine a round fruit—just its image, painted on a stone wall. And I imagine a ray of energy, traveling through the stone and emerging from the stone to fly over the whole city. This helps. This imagining this thing helps.

What does it help?

(Pause)

It helps me. If I try to say what it helps—me—that separates me from myself and that does not help. So I do not explain why it helps, even to myself. I just say and know, it helps. Which is much like being in, or traveling towards, the Hotel Radio. Just remembering it, even from inside one of its rooms, and I don't know if there are many such rooms or only a few—but it helps.

It helps.

Hello, this is a part of the hotel radio, and it helps.

Self-discovery in a hotel? This does not seem possible.

(Bring in a radio)

What's this?

(Pause)

Wait a minute—do I want a radio in my room? or will that make it difficult to know whether or not it's me doing the talking.

Where shall I put it.

I'd rather have my suitcases delivered than a radio.

Soon.

When.

Soon

When is soon.

The radio could take your mind off your problems.

Is that true?

Sometimes.

Plug it in but don't turn it on.

No. You plug it in.

(Pause)

Just put it on the table.

(Done, gone)

Self-discovery in a hotel. This does not seem possible.

LINGUA FRANCA

written, produced and directed by

lou mallozzi

A man stands on the shore,
on a strange beach.
He has never been here before,
and everything is fresh and crisp in its unfamiliarity.
Everything is abnormally vivid.
He feels the wind long the edges of his ears,
and he realizes that this is surely the same wind as in his homeland,
on familiar beaches in safe harbors,
but he has never noticed it touching his ears.

He names this new wind "ear-edge-breeze."

He knows that it is his noticing that makes this breeze unique,
different from the winds of his past and his home,
and that his noticing means that the wind gets a new name,
and,
further,
that the wind is changed by his naming it,
his noticing and acting on this noticing.
This all happens in an instant.

His eyes blink and shift slightly to the left and slightly down.
He sees the tendril of a vine that curls in a corkscrew shape that is precisely
the same as the shape of the wood shavings that covered the floor
of his grandfather's woodworking shop.
As his grandfather planes the edge of a mahogany table top,
the golden corkscrew-shaped shavings grow out of
the plane's top and fall to the floor.
As a boy,
he would focus on each one and they would pile up as
the morning wore on.
Now, on this strange beach,
the tendril of this vine

EH, AND, EH, WE HAVE TO MAKE A PARENTHESIS TO MAKE SOME CORRECTIONS. IN PAGE NUMBER 4, YOU HAVE NUMBER 1, 2, 3, 4, AND 5, ALL RIGHT? YOU HAVE TO CROSS OUT NUMBER 2. CROSS NUMBER 3. INSTEAD OF NUMBER 4 WRITE NUMBER 2. INSTEAD OF NUMBER 5 WRITE

NUMBER, NUMBER 3.

INSTEAD OF — GO TO

PAGE 5 — INSTEAD OF

NUMBER 6 YOU WRITE

4. INSTEAD OF

NUMBER 7 YOU WRITE

5. INSTEAD OF NUM-

BER 6 — EH, 8 — YOU

WRITE 6. PAGE NUM-

BER 5, POINT 8, THERE

SHOULD BE 6, ADDEN-

DUM, HM? AND YOU

GO INTO THE SECOND

LINE, YOU WRITE

— solana dulcamara —
which he has surely seen before on familiar homeland shores
— he notices that the curve,
pitch,
and thickness of the tendril are the same as those of the wood shavings,
only the color is different.
He has never noticed this exact relationship before.

He names this plant "grandfather-work-vine."

solana dulcamara　　nightshade　　bittersweet nightshade

As for my eyesight, it is only useful for navi-
S F R M S GHT, T S NL S F L F R N V
gating in the dark. At night, I can find my way
G T NG N TH D RK. T N GHT, C N F ND M W
by the stars. My left eye squints as my right eye
B TH ST RS. M L FT SQ NTS S M R GHT
intersects the horizon by means of the brass sex-
NT RS CTS TH H R Z N B M NS F TH BR SS S X
tant, which must be calibrated exactly, otherwise
T NT, WH CH M ST B C L BR T D X CTL , TH RW S
I will miss the mark by hundreds of miles. We'll
W LL M SS TH M RK B H NDR DS F M L S. W LL
be aimlessly adrift for weeks.
B ML SSL DR FT F R W KS.

Failing that, I can go by sense of smell — the
F L NG TH T, C N G B S NS F SM LL — TH
open sea has a constant cool salty smell with
P N S H S C NST NT C L S LT SM LL W TH
occasional waves of sour warmth. In this consis-
CC S NL WV S F S R W RMTH, N TH S C NS S
tant atmosphere, it should be easy to detect the
T NT TM SPH R , T SH LD B S T D T CT T
smell of distant land — the dry metallic smell of
SM LL F D ST NT L ND - TH DR M T LL C SM LL F
sand, the camphorous and acidic smells of plants,
S ND, TH C MPH R S ND C D C SM LLS F PL NTS
the dank and humid smells of animals.
TH D NK ND H M D SM LLS F N M LS,

As a last resort, I can proceed by my sense of
S L ST R S RT, C N PR C D B M S NS F
hearing. The sounds of the sea change as one
H R NG, TH S NDS F TH S CH NG S N
approaches land. In the open sea, the waves lap
PPR CH S L ND, , N TH P N S TH W V S L P
against the ship's hull in straight waltz time.
G NST TH S P'S H LL N STR GHT W LTZ T M.
But, nearer to land — even before it is visible —
B T, N R R T L ND - V N B F R T S V S BL
the rhythm becomes irregular. First three beats,
TH RH THM B C M S RR G L R, F RST THR B TS,
then four, then perhaps five, three, four, two —
TH N F R, TH N P RH PS F V , THR F R TW -
as if the sea were stumbling over itself to
S F TH S WR ST MBL NG V R TS LF T
escape the land.
SC P TH L ND,

$_o$s f$_o$r m$_o$ $_o$s$_o$ght, $_o$t
$_o$t n$_o$ght, $_o$ c$_o$n f$_o$nd
sq$_o$nts $_o$s m$_o$ r$_o$ght
s$_o$xt$_o$nt, wh$_o$ch m$_o$st
w$_o$'ll m$_o$ss th$_o$ m$_o$rk
w$_o$ks.

F$_o$l$_o$ng th$_o$t, $_o$ c$_o$n g$_o$
h$_o$s $_o$ c$_o$nst$_o$nt c$_o$l s$_o$lt
th$_o$s c$_o$ns$_o$st$_o$nt $_o$tm$_o$s
$_o$s$_o$ t$_o$ d$_o$t$_o$ct th$_o$ sm
s$_o$nd, th$_o$ c$_o$mph$_o$r$_o$s
d$_o$nk $_o$nd h$_o$m$_o$d sm

$_o$s $_o$ l$_o$st r$_o$s$_o$rt, $_o$ c$_o$n
Th$_o$ s$_o$nds $_o$f th$_o$ s$_o$
$_o$p$_o$n s$_o$, th$_o$ w$_o$v s l$_o$p
w$_o$ltz t$_o$m . B$_o$t, n$_o$r
rh$_o$thm b$_o$c$_o$m s $_o$rr$_o$
th$_o$n p$_o$rh$_o$ps f$_o$v , th
$_o$ts$_o$lf t$_o$ $_o$sc$_o$p th$_o$ l$_o$n

COMMA, AND THEN YOU CONTINUE IN THE FOLLOWING LINE "THEIR PROJECTION." THEN, WHERE IT SAYS B YOU HAVE TO ADD THE CONJUNCTION "AND," AND NOT PERIOD, BUT COMMA, "BOTH WITH EXTENSIVE." AND SINCE WE ARE COR-RECTING THINGS THAT ARE INCORRECT, MOVE INTO PAGE NUMBER 10.

bittersweet herb blue nightshade felonw

PAGE NUMBER 10 HAS POINT 1, 2, AND THEN 3. CROSS IT OUT AND PUT A SMALL E. AND THEN YOU CROSS OUT THE SMALL A, B, AND C, AND D, AND THEN YOU CROSS OUT A, B, AND C, AND YOU CHANGE D FOR A SMALL F. IS THIS CLEAR? WE ARE IN PAGE 10. NOW, AFTER F, IT COMES NUMBER 3. NOW YOU HAVE TO WRITE NUMBER 3 AND CROSS NUMBER 4.

ng $_o$n th$_o$ d$_o$rk.
't $_o$
m$_o$ns $_o$f th$_o$ br$_o$ss
v$_o$s
ll b$_o$ $_o$ml$_o$sl$_o$ $_o$dr$_o$ft f$_o$r

As a mein eyesight, essa ist only utile zum navigating alla bei dark. Alla Abend, I posso finden my via durch the stelle. Mein left occhio blinzelt as mio recht eye interseca der horizon per Mittel of il Messing sextant, che muß be calibrato genau, otherwise io werde miss il Kern by centinaia von miles. Noi sein aimlessly deriva für weeks.

$_o$n s$_o$
s $_o$f s$_o$r w$_o$rmth. $_o$n
dr$_o$ m$_o$t$_o$ll$_o$c sm$_o$ll $_o$f
ts, th$_o$

Failing quello, ich can andare zu sense di Geruch — the aperta See has un beständig cool salato Geruch with occasionale Wogen of agro Wärme. In questa fest atmosphere, deve sein easy scoprire der smell di fern land — il trocken metallic odore nach sand, i kampferisch and aceto Geruches of pianti, di dank e feucht smells di Tieres.

ng.
d. $_o$n th$_o$
tr$_o$ght
s v$_o$s$_o$bl — th$_o$
f$_o$r,
v$_o$r st$_o$mbl$_o$ng $_o$v$_o$r

As una letzt resort, io kann proceed al mein sense di Gehör. The suoni von the mare ändern as si näherkommt land. Al die open mare, die waves lambiscono gegen the barca Rumpf in dritto Waltzer time. Ma, nahe to terra — einmal before essa ist visible — il Rhythmus becomes irregolare. Erste three battiti, dann four, poi vielleicht five, tre, vier, two — come ob the mare wäre stumbling sopra sich to sfuggire
das land.

garden nightshade scarlet berry staff vine violet bloom

I could, then, eventually find my way, unless, like Odysseus,
I was bound to the ship's mast with my ears plugged with wax,
my eyes blindfolded, my nose stuffed with cotton,
as the seaspray and wind tear at my forehead and shoulders.
This would put me at the mercy of a poet,
who would be obliged to find a loose thread in the fabric
of the drama, so that some tragedy
would unravel the story, inviting disaster,
mayhem, and death, all at the service
of my heroism which would expunge years
of evil deeds, and would serve as fitting
retribution against my pitiable enemies — who
are numerous and international — and
would culminate in exultation for me;
and I would surely receive riches rivaling those of the pope,
and be subject to the erotic attentions of the land's
most beautiful women,
whose fingers would caress the many scars on my body
that are my immortal
pink badges of courage,
perseverance,
and guile.

ALL RIGHT? NOW GO

TO PAGE NUMBER 11.

PAGE NUMBER 11,

NUMBER 1 AND 2, A —

SMALL A — IS LEFT

THERE. THEN YOU

CROSS B, C, D, E, AND

F, CROSS IT OUT.

EVERYTHING IS RELATED

TO A. TURN TO PAGE

12. CROSS OUT THE G

AND PUT A SMALL B.

THEN GOING TO PAGE

12, GO TO CAPITAL D

— CAPITAL D. POINT NUMBER 1 IS ALRIGHT. CROSS OUT POINT NUMBER 2, BECAUSE EVERYTHING IS RELATED TO NUMBER 1. AND POINT NUMBER 3 SHOULD BE SUBSTITUTED BY 2. PAGE NUMBER 13. CROSS NUMBER 4 AND 5. AND FINALLY PAGE NUMBER 14. NUMBER 1 IS ALL RIGHT, CROSS NUMBER 2, 3, 4, 5, 6, AND 7. EVERYTHING — OFF.

Lingua Franca is an experimental narrative exploring navigation, both as the subject of the text and as a metaphor for listening. The utterance of language has a number of forms: description, instruction, repetition, phonemic fragmentation, multilingual translation, and melodrama. In addition, language and narrative are constantly dislocated and relocated into a variety of sonic spaces.

The 27-minute radio work was written, produced, and directed by Lou Mallozzi. It was commissioned by New Radio and Performing Arts for New American Radio in 1992. The narrators are Claudia Renchy, Monika Bruder, Larry Bull, Lauri Macklin, Gian Luca Ferme, and Lou Mallozzi. The musicians are Max Callahan and Michael Zerang.

This text was designed by Kali Nikitas.

Around **Naxos**

a radio "film"...[1]

Kaye Mortley

a patient labyrinth of forms, his own portrait
—borgès, on velasquez

The curator of Greek, Roman, and Etruscan antiquities at the Louvre is distracted. He has a lot of other things to do. The Naxos project is almost completed, but I have asked him to "read" the myth of Ariadne for me... as it is written on those black-and-red Greek vases of which he is a specialist.

. He clears his throat. **Pages of a book are turned.**

Footsteps. **Floorboards creak.** **A door closes.**

1.

You remember the myth...?

Theseus goes to Crete to kill the minotaur.

He meets Ariadne, the daughter of Minos...

and she falls in love with him.

Ariadne gives Theseus a skein of red wool, so he can find his way out of the labyrinth.

We meet early one winter morning in this office.
There are a lot of books with photos of vases on his desk.
The floorboards creak.
The sound engineer directs his Schoeps microphones at the curator, and thus

towards the window behind him, and the roar of traffic along the Quai du Louvre...
This was a mistake.

. Steps on stone.   A heavy door slides closed.   Keys.

2.

It was with the help of Ariadne that Theseus was able to leave Crete.

He carries her off on his boat.

And then the boat stops

at Naxos...

put the past in the present. the present is magic
—bresson

Yes.
You do remember the myth.
But you don't know from where...

Was it that red book of Greek and Roman myths you had in your first year in high school?
Or was it perhaps the Argonauts, *that Australian radio program for children, which supplied so much of your cultural prehistory...*

However
you do not remember that Pasiphaë was enamoured of a bull, and that the minotaur (the issue of this hapless passion) was, thus, Ariadne's half-brother. Nor do you recall a skein of red wool. *For you, it was a ball of brown string, like they had at the grocer's, or the post office. And your labyrinth was reinforced concrete corridors (like at the swimming pool in the country); or else a privet maze (like the one in the park)*
with an empty space
at the centre.
As though the minotaur,
who is perhaps the real
centre of this story/
labyrinth
had always been
missing.

bring together things which have never come together before and which it seemed unlikely would ever meet
—bresson

Yes.
You remember the myth.
But in the same way you remember something you have recorded.
You live with it, listen to it, touch it, caress it, work on it until it becomes
the memory of the memory
of itself
and
the forgetting
of the first recording.

some people start with a documentary and end up with fiction (flaherty); others start with fiction and end up with a documentary (eisenstein, *que viva mexico*)
—godard

. ***Female voice, reading the words of a song, in Greek.***

. ***Port atmosphere, fading up slowly.***

. ***An instrument, tuning.***

. ***Greek music relayed through acid loudspeakers.***

. ***Then: a man, singing.***

The boat arrived late.
They are often late, these boats.
It always takes so long to go from Piraeus, to an island... the wind, the sea, or something else...
Time stands still.
You begin to understand the cosmogonical attitude of Homer, and that the Mediterranean is indeed the centre of the world.

The boat was late.
The hotel on the port was full.
(The guide book had warned that tourist accommodation was scarce.)
Someone suggested Maria's, a new hotel, half-finished, up a dirt track, out of town...

my bag seems heavy: a couple of pairs of shorts
a swimming costume
a Sony Walkman professional cassette recorder
a small mike
headphones
12 x 90'00" chrome cassettes...

Maria rented out plastic-covered sofas under glaring neon lights in her lobby, for what was left of the night.

The next day, on the square, down by the port,
you look for a real place to stay.
Bad loudspeakers bleat out music you thought
was bouzouki.

(. Song continues, under.)

3.

Like a tempest,

our encounter.

Shipwrecked.

I was shipwrecked

in the sea.

Your eyes are the sea.

(. Song, cross-fading with sea.)

Broken, broken my heart.

Your eyes have broken my heart.

(. Sea.

. Siren.

. Siren.

. Siren.

. Siren.)

Much later, K.T. comes to a studio to translate what was on the tapes I had recorded. Her mother is from Naxos, but I'd forgotten this when I called her. She explains that the songs are love songs, peculiar to the island. The songs of estranged lovers.
Songs sung into space
- into the airwaves -
to someone
somewhere
out there
listening -
or not
wanting to hear -
or not.

Radio/songs, these love/songs

no accompanying music to prop up, reinforce. no music at all. (except, of course, when played by instruments you can see.)

the sounds must become music
—bresson

. The sea, closer.

. Monteverdi's* Arianna *(Emma Kirkby) : "lasciatemi morire..."

In a library at Radio-France, I light upon a music review (*la Revue musicale*) which informs me that Monteverdi's *Arianna* dates from round the same time as the death of the composer's wife, and is thus a poignantly ill-timed commission... this *divertissement* destined to celebrate the marriage of François de Gonzague and the Infanta Marguerite de Savoie...

4.

music alters, even destroys one's perception of reality, like alcohol, or drugs.
—bresson

Lasciatemi morire

Lasciatemi morire

e che volete voi che mi conforte

in cosi dura sorte

in cosi grand martire

lasciatemi morire...

Of this nuptial celebration, the lament remains...
most of the rest is lost.
Like a radio program.
Most of which is always "lost."
The only trace it leaves is in the mind/the ear/the mind's ear of the listener.
And who can say what that will be?
What I think I say
is not
what you hear.
And then
there is always
that empty space
the free zone
the centre of the labyrinth
where you
the listener
can
write read weave paint hear
your own story.

(. Sea... Monteverdi.)

5. Enter Apollo.

(. Seagulls, off.)

Venus implores Eros to save Ariadne, whom Theseus would abandon on a deserted shore.

(.* Arianna: *text, read in Italian.)

Theseus arrives at Naxos with his lover.

(. Seagulls.)

Night falls.

(. Sea closer. Disc: "o Teseo...". The text is read in Italian; fades into the sea.)

Theseus would allay Ariadne's fears.

And, perhaps, you will not hear very much of what I consider to be my "story."

. Sirens: 1, 2, 3 mixing with alpine horn (Pierre Mariétan).

Sea out. ***Alpine horns continue, fading.*** ***Silence.***

Another winter day
late afternoon, paris-grey.
I was tired of cutting tape.

edit while you are shooting (recording). this will give you kernels (of *strength*, of security) for the rest to latch on to.
—bresson

Question: do I like cutting tape?

Answer: more than digital editing, always virtual, never irrevocable, not like real life...

Oh I'm sailin' away My own true love,
I'm sailing away in the morning.
Is there something I can send you from across the sea,
From the place that I'll be landing?

...but it is tiresome...
except for the pleasure of trying to create small scenes, tiny worlds which, without *being* real, contain something real—fragments of the moment when the angel passed...
Radio is about angels passing.
You write the passing of the angel with your scissors. Sometimes feathers fall into the wastepaper basket.
Feathers that were beautiful,
and which you loved.

build on
emptiness,
silence,
immobility.
—bresson

. Cicadas.

00'05": a crowd, animated, fading up.

00'10": a Greek band, tuning.

00'19": it starts to play.

00'33": Dylan's harmonica just audible in the Greek music.

00'46": the harmonica cuts through the village scene which, however, continues under ("I don't understand what you want to do": the sound engineer, when mixing) as in a dream, or some space of memory.

01'14": the harmonica continues.

01'26": Bob Dylan (in a Tom Waits mode) starts to sing. etc. etc.

6.

Not a word of good-bye,

Not even a note.

She'd gone with the man

In the long black coat.

I was tired of cutting tape so
I went to the supermarket.
That's where I found the song.
It wasn't new.
But I didn't know it.
The refrain reminded me of
"Boots of Spanish Leather"
and something Professor C.R.
had said when I was making that
extremely long Bob Dylan program with him for the A.B.C.,
so long ago...

... namely, that "Boots of Spanish Leather"
was really about a woman leaving a man
but pretending that it was the other way round.

Perhaps Theseus did not leave Ariadne.
Perhaps Ariadne left Theseus
having glimpsed Dionysos on the beach
("the old dance hall on the outskirts of town")...
Perhaps the whole Ariadne story was the result of male-written, male-dominated mythology...
But in a sense, did that really matter, except as another angle from which to approach the myth?
What seemed to matter more (along the same extrapolatory lines of thinking) was that Theseus was a hero; heroes become history (the past). And that Dionysos was a god (immortal) but a god associated with the pleasures of the senses (the present).

Could Dionysos have something to do with radio?

A radio work exists only in the present,
i.e., at the moment when a tape is "read"
by a tape recorder.

On Naxos is about time.
Radio has to do with time.

"How come you are so interested in Bob Dylan?"
I finally dared to ask, after recording Professor C.R. for two days.

"Oh," he said.
And he looked out the window of that café close to Radio-France.
"Bob Dylan always tends to get mixed up with one's life,
and one's loves."

On Naxos is also about love.
Radio often has to do with love.

. Cicadas.	*Sirens: 1, 2, 3, 4.*	*Cicadas fading.*	*Siren.*
Cicadas out.	*Silence.*		

your imagination should be concerned less with events than feelings, which you should strive to keep as documentary as possible.
—bresson

Every day on Naxos
I took my Sony Walkman professional
my headphones
and my small microphone
everywhere I went...
to cafés, to the beach, on the bus, on the ferry.
I often hated my tape recorder.
But I had decided to record
Naxos
&
myself discovering Naxos

the city is an ideogram
the text continues
—barthes

& the never-ending text generated by Naxos

where does the writing begin?
where does the painting begin?
—barthes

Perhaps the recording was the writing, and the mixing would be the painting,
I thought, as I walked, microphone extended greedily,
randomly towards every sound.
But I knew that
the writing & the painting/
the recording and the mixing
were all of a piece.
And that I would have to find some way of using this fairly basic equipment to advantage, to avoid the straggly, ill-defined sort of sound which would suggest that I (the first listener) and the sound (destined to a second listener) were light-years apart. I found that if I focused on some irrelevant but precise acoustical detail—footsteps, a voice, a child, a door opening—then, the rest of the "scene" seemed to fall into place. And the same way that the preverbal, the nonverbal (a breath, a sigh, a hesitation) can become the subtext of an interview, here these small irrelevant details seemed to become the unsaid story of what I was recording.

Except that, here, there was no "story."
Just the immanent, brooding subject of Naxos.
An island of mind.

***à bout de souffle* is just a story, not a subject. the subject is something simple and vast (revenge, pleasure) that you can sum up in twenty seconds... the story would take twenty minutes.**
***le petit soldat* is a subject...**
—godard

No real story.
Just a place.

I decided to ask seven male writers, of various nationalities,
to generate fragments of stories.
Each was asked to write a postcard, addressed to a woman,
as from Naxos.
Perhaps Theseus had never left...? perhaps he had returned...?
perhaps these missives were not from Theseus...?
In any case, before receiving the postcards, I decided that all would be read by the same male voice.
Some "postcards" arrived quickly; others came in slowly, over a year,
or more, from
Harry Mathews, Hugo Santiago, Sévéro Sarduy, Jean-Louis Schéfer, Thomas Shapcott, Jean Thibaudeau, John Tranter...
People played by the rules, or not.
Anyway, there were no rules for this game...

7.

She's got a tape recorder... can't you see?

. My dear friend,

It occurs to me that the name of the largest island

in the Cyclades probably has an unlikely terminology:

Naxos must come from *nao*, meaning: "whey overflowing"...

all the poetry of the island reeks of goat cheese...

. Don't you see...she's got a tape recorder...?

...no facts, facts are always a lost cause; but ways of saying, ways of doing...
—françois niney on marcel ophuls

Among the classic sites for gathering urban text:
the market.

(The market. People buying, selling, chatting, negotiating, wrapping.)

8.

How much does this weigh?

The grapes are good...two kilos for 350 drachma.

There are tomatoes over there...good tomatoes for salads.

(A moped goes by.)

. Will I still see you in Lausanne, in two weeks?

You hope to be invisible
but you are not.
Floating in the air are voices
which you hope to steal, unnoticed
but cannot.

A visible microphone changes the ecology of any human situation.
An invisible microphone can "take" quite freely.
But how interesting is this?
A microphone is not just a spy, or a thief.
A microphone is also an interlocutor.
There can be a two-way exchange
even if you (the mike-stand, as it were)
never open your mouth.

9.

But she's stealing our voices...she's recording everything we say...that's a tape recorder.

The first concept had been to make a series of acoustic postcards to be sent (broadcast) as from Naxos.

the microphone (...)
might have frozen time
encapsulated memory
on a tape
to return to
to relive
and

. 00'00": in Greek,
a female voice reads a love song of Naxos.
. 00'19": the same song, sung unaccompanied, by the same voice.
. 00'38": a door slams, footsteps.
. 00'54": Catullus.

But time passed. And then some more. The 12–18 hours of tape were first cut to discard, then to shape; and then they were deciphered (translated).

No.
Naxos was not going to be a series of acoustic postcards.
Rather a skein of everyday life, of different colours and textures, unraveling with knots, and breaks.

. 00'00": a window opens.

. 00'18": outside, the sound of children playing in the distance.

. 00'30": the window closes.

. 00'43": a door opens—a flurry of birds' wings.

. 01'05": a dog scampers by.

. 01'15": a gate opens, the chain rattles.

. 01'45": footsteps in leaves.

. 02'06": men walk by, chatting in Greek.

. 02'23": a donkey... etc. etc.

And, like broken pieces of those black-and-red vases from the Louvre, fragments of many Ariadne stories would appear, disappear, in the unraveling skein of sound. Some sort of almost random collaged text was coming together...

-Catullus
-Monteverdi / Rinuccini
-the love songs of the island
-the postcards
-a collage of cards people had sent me
-Bob Dylan's song (translated)
-Tom Waits's song (not translated) ...

10.

A month of change,

relief,

instability.

I am in another world.

The Arafura Sea

red roads

anthills...

And you are in another world...

Different voices/various accents/several languages (Greek, French, Italian, Latin, English) circulating...

A fragmentary incantation.
A palimpsest.
The story of Ariadne rewritten in off-camera voices.

radio is an off-camera universe.
—butor

a. The protagonists are invisible...
Everything is out of the range of the camera.
b. There is no camera.
a. In radio, everything is happening somewhere else...
b. Where...?
a. Somewhere invisible...
Radio is such stuff as dreams are made of.
Or memory...

. A group of children—Dimitri, Anna, Sophia—in the street.

They are selling pebbles on which they have painted:

"from Naxos, with love..."

radio, the acousmeter...please forgive a truism:
radio, by its very nature, is acousmatic.
—chion

Unlike the cinema, where the off-camera voice signals an absence, a separation of the voice (which you hear) from a body (which you cannot see), the voice in radio is not disincarnate: it is its own body.

not actors: models. models work from the outside, in.
actors work from the inside, out.
—bresson

Radio voices must be "real" voices.
Not voices pretending to emanate from some body which, by aesthetic accident or design, we do not see.

In *Naxos*, the voices chosen were:
two Italian teachers of Italian
one production secretary
one Greek multimedia artist
one American performance artist
one American-living-in-Paris
five university students
&
two actors—
one had worked with Peter Brook
the other, with Antoine Vitez.

11.

Enter a messenger.

He saw the boats leaving...

saw her watching them go...

saw her run down into the sea

begging the waves to carry her off...

But it is the last voice, the curator (in a sense the only "real" voice in the piece), which is the acousmatic voice *par excellence.* As in the pythagorean sect, this is the voice of the master behind the curtain, dispensing knowledge to his disciples. "Behind the curtain" there is the museum. A museum is a place of knowledge and thus, of power. This off-camera voice is not I, we, you. It is not a voice in which we hear our own body, pulsing, vibrating.
It is a voice that informs.
A voice documenting reality.
A documentary voice.
A voice telling us that reality/
is transparent.

12.

...but there are other images, like this one here, where the goddess Athena would seem to be ordering Theseus to leave...he seems to be making a gesture of regret...Ariadne is expressing sorrow...but she will immediately be picked up by Dionysos, who seems to be hanging around, waiting to take over where Theseus left off...

. 00'00": a piece of paper (close-miked) is slowly torn up.

. 00'02": horses' hooves, sleigh bells approaching.

. 00'25": vying with the sleigh, Tom Waits's, "Tom Traubert's Blues."

. 00'33": the words surface...

. 01'15": are overtaken by the sleigh.

. 01'38": the horses fade, slowly.

Is reality ever transparent?

the moment you perceive reality, it is no longer real.
—bresson

Is reality ever transparent?

(. Horses, still fading.)

13.

Leave her alone Kostas...leave her alone!

We'll go down to the port, then we'll come back...

tell your mother to meet us there.

(. The street. A child wails.)

. Anna! no no no...!

The sound which conveys, creates, re-creates "reality" is often transparent.
It does not exist, as sound.
You can pass—untrammeled, unimpeded—through the sound to
the message
the truth
the story.
The sound does not solicit you
it does not *bother* you
it does not tug at your sleeve
pull at your coattails.
It does not oblige you to stay there—
inside the sound—
and to listen to
the language that it speaks.

A language made
of colours and shapes
of silence
of rhythms
of time.

. 00'00": a child crying in the street, footsteps.

. 00'09": a glass-seller goes by calling his wares.

. 00'18": "The Green Green Grass of Home," over a radio in a café.

. 00'44": a donkey.

. 01'15": the song ends; the glass-seller continues.

Naxos, then, would be based on real sound.
The sound of "real" life.
Documentary sound, as it were.
But sound—
and the texts & the voices reading them—are part of the sound;
used in such a way that
it is the message
　　the only "story."

14.

Georges, listen...do you hear...who's that?

. They say that she cried out loud...

That she climbed the steep cliffs, her eyes scanning the endless sea...

that she ran into the waves,

almost drowning.

. The street: birds, children playing.

. A chair mender goes by calling: "kariklas."

. Children close; footsteps.

. Footsteps climbing stone stairs.

. Women's voices, present; birds louder.

. Bouzouki, over a radio in a café mixing with a love song of Naxos.

. Over the love song: Catullus (Ariadne's recrimination).

. Monteverdi (Emma Kirkby): "O Teseo. . ." etc. etc.

raw reality by itself never makes for truth.
—bresson

Note

1. *on naxos* (*à naxos*)—originally produced for the atelier de création radiophonique, france-culture, and broadcast 23 february 1993—was mixed by monique burguière, assisted by bruno roncière. duration: 2h00. the voices referred to in this essay are those of this first production: steve gadler, jani gastaldi, françois marthouret, rita rapaport, christine rey, domenico romeo, stuart sherman, katerina thomadaki & elise bensa, marion degorce-dumas, catherine hass, anna mortley, sarah schwarz. *on naxos* has since been broadcast in a 60'00" version by: A.B.C, Y.L.E., and S.R.

   the texts numbered 1-14 refer to fragments of the text used in the adaptation:
   1, 2, 12: the curator
   3, 7, 8, 9, 13, 14: the street—"conversation poems"
   4, 5, 11: rinuccini/monteverdi
   6: bob dylan
   7: jean-louis schéfer
   10: kaye mortley
   14: catullus

quotations

barthes, roland
1989 *empire of signs*. new york: noonday press.

bresson, robert
1975 *notes sur le cinématographe*. paris: gallimard.

butor, michel
1992 interview with author. paris.

chion, michel
1993 *la voix au cinéma.* paris: éditions de l'étoile/cahiers du cinéma.

dylan, bob
1964 "boots of spanish leather." *the times they are a-changin'* LP. columbia.

godard, jean-luc
1985 *les années karina.* paris: flammarion.
1985 *les années mao.* paris: flammarion.

niney, françois
1994 "l'histoire peut-elle se répéter." *images documentaires* 18/19.

Music to the "nth" Degree

Brandon LaBelle

Noise: 1. (a) a loud or confused shouting; din of voices; clamor (b) any loud, discordant, or disagreeable sound or sounds.

Music: 1. the art and science of combining vocal or instrumental sounds or tones in varying melody, harmony, rhythm, and timbre, esp. so as to form structurally complete and emotionally expressive compositions.

—Webster's New World Dictionary of the American Language, Second College Edition

Noise as a sudden aural disturbance ruptures the coherency of musical composition. It makes its sonic appearance as a discordant irritation of musical resolution. Interfering with harmony, displacing the moment of calm, Noise prolongs disquietude by opening up the divide between crisis and restoration, certainty and uncertainty.

The French economist and political advisor Jacques Attali, in his book *Noise: The Political Economy of Music* (1985), theorizes that noise functions not only as a musical construct, an antithetical necessity, but as a greater force within the social order by announcing the chaotic fraying of its governing codes. Noise attacks the status quo, the norms that govern relations and dictate one's position as an individual. In essence, it embodies that which disturbs the strata of social relations.

Attali characterizes Noise in terms of the "carnivalesque": an excess of expenditure, a rupture of the rational, a lapse in the coherency of social behavior. For Attali, Noise functions as a release of primal energies, a kind of ritualistic enactment of disorder and violence. As an example, he examines the relationship between Lent and Carnival as depicted in Breugel's painting from 1559, *Carnival's Quarrel with Lent*. Preceding Lent, Carnival was a period of indulgence where one surrendered to the appetites and their excessive fulfillment, allowing one to stray from the social order, escape class distinction and economic constraint. It also allowed one to safely stray from the limits of Christian etiquette, without repercussion or consequence. What Attali rightly points out is that Carnival functioned within the social order to allow individuals an escape from its constraint, if only to return more devotedly to a pious lifestyle through the self-sacrifice of Lent. In essence, Carnival reinforced the authority of religion by dictating when and how this disorder could occur.

Attali uses Breugel's painting as a metaphorical rendering of Noise in order to point out how music is bound up within struggles of power. Attali "hears" within the painting the moment when religious order confronts its own subversion and the inherent disorder of the carnivalesque unfolds in spite of social limits. It is not so much the depiction of any musical event, though music is present. It is more a "visual harangue" that Attali perceives in the complexities of the crowd, the intermingling of bodies, the presentation of the hierarchy of class as found within the gestures and articulations of actions: jugglers, nuns, cripples, workers, drunks, and dirty children, all come together in the painting's perspectival point which overwhelms the visual frame.

Following Attali, the "carnivalesque" appears as a ritualistic enactment of murder and self-destruction. It unfolds as theatre, establishes its own order, its own ceremonial practices, from parades to debauched orgies, drinking contests to extravagant costumes. Yet its ceremonial practices bring Carnival into the social order; it follows the design of a belief system, however unstable. Yet through this "enacting" one becomes involved. One becomes immersed in the theatrics. As a participant in this ritual one gets caught up, slips into the role a little too fully, too extravagantly. The carnivalesque spins in on itself, leading to a dead center of its own destruction. It implodes. In this way, the carnivalesque is a threat to every order, for it brushes against uncontrollability—it flirts with the beyond. For this reason, though, it is allowed to occur. The carnivalesque is given its moment, and through the enactment of violence the social order is allowed its own dissolution, and ultimate restoration.

Music functions according to a certain order. It is heard as music because it follows a logic that is comprehensible and that refers to a past. As a cultural institution it has authority. The musical resolve, which we can define as the reestablishment of harmony within musical composition—resolution of a tonal tension—functions as part of this authority. It is a kind of reassurance that things are in their place and that nothing is askew: it concludes sonic suspense. Yet the resolve needs its crisis, its antithesis, if only, in the end, to put things right.

Music has within it its own beyond, a threshold to chaos, an otherworldly logic that defies the rules: it strays in the joining of notes, slips into a fold of improvisation. Noise appears, reveals itself inside the space where the hand touches a string, where the mind imagines how something will sound. As the antithesis to composition, Noise forces a musical logic to rethink its own boundaries.

In his writings, the Japanese composer Toru Takemitsu discusses the differences between Western traditional music and traditional music of Japan. One of these differences, he observes, is found in the relationship to "noise":

> We can see that the Japanese and Western approaches to music are quite different. We speak of essential elements in Western music—rhythm, melody, and harmony. Japanese music considers the quality of sound rather than melody. The inclusion in music of a natural noise, such as the sound of a cicada, symbolizes the development of the Japanese appreciation of complex sounds. (Takemitsu 1995: 65)

This appreciation can be seen in Eastern music in general, which is based upon a wider set of harmonic rules than Western classical music, and is generally more involved in the details of sound as discovered within single instruments and performances. Noise exists not in opposition to this attitude but rather is seen as the result of a musical pursuit, a positive by-product. In contrast, the Western classical tradition bases itself on harmonic rules that tend to refuse these sonic details. From here, it is easier to understand why "Noise-Music" comprises such a provocative body of work in Japan. Though Takemitsu was discussing traditional Japanese music, such as *gagaku* (Japanese court music) it is interesting to consider his observations in relation to contemporary Noise-Music.

The radicality of the very term Noise-Music—placing the two words against each other, if only to suggest an electrical horizon of musical experience, a magical coupling—imprints itself onto the imaginary through its own voltage, its commitment to prolonging suspense and drifting farther away from a conclusive shore.

Following British and American 1960s psychedelia (Hawkwind, Black Sabbath, Jimi Hendrix, etc.), German experimental and electronic music of the '60s and '70s (Can, Cluster, Kraftwerk), and the "no-wave" scene in New York in the late '70s (DNA, Teenage Jesus and the Jerks), Japanese Noise-Music of the past 20 years is characterized by an extreme use of electronics to build up sonic walls of feedback, electronic loops, and extreme levels of volume, assaulting not only the ear but the entire body. Noise-Music aims to traumatize—to disrupt the limits. Early Noise bands, such as Hijokaidan, played a kind of improvisatory punk—spastic, cut up, and dysfunctional, bordering more on performance art than punk music, a kind of "punk-happening" of amplified vomiting, glossalalic utterances, broken instruments. Others such as Merzbow, KK Null, and Keiji Haino—ambassadors of what is called "Extreme Noise"—use guitars and homemade electronics to produce steady washes of electrical noise. The systematic layers of Glenn Branca multiply infinitely: music to the "nth" degree.[1]

On whichever side of Noise-Music one stands, it appears as a compelling extension and confluence of Cage and fluxus performance and the rock psychedelia and punk traditions. It takes the boldness of punk, the improvisatory spirit of psychedelia, and the ethos of noise as a possible music—which Cage and Fluxus brought to the forefront of avantgarde practice—and appropriates all of this in a kind of living theatre of subcultural extremity.

These artists, along with many others in Japan (such as CCCC, Kyoshi Mizutani, Masonna, MSBR, Aube, Solmania, K2), expand Noise to a larger scale. Its proportions increase, its fevers become more heated, the volume louder, the performance more intense. Noise-Music widens the void, supporting a kind of sonic violence that unhinges the possibility of the musical resolve, by situating itself against the moment when harmony must reappear. Yet is Noise-Music, like Carnival, a moment of chaos governed by the directives of the order it seeks to escape, to push aside?

In *Japan: A Reinterpretation* (1997), Patrick Smith investigates the United States' reformation of Japan following WWII. He identifies a hidden dissatisfaction in Japanese culture that began with the sudden change in American policy during the Occupation.

Just prior to the cold war, the Occupation sought above all to establish democracy within Japan and to install social policies that would support democratic growth and undermine the imperialistic rule of the emperor. Yet, with the sudden emergence of the Soviet occupation of Eastern Europe, which spread throughout Asia from North Korea and China to Vietnam, and the beginning of the cold war, the United States shifted its approach to Japan by "reversing" the initial democratization, supporting instead a return to prewar politics. Installing leaders of the war into high-ranking positions, and casting Japan as a docile ally of the U.S. and the cold war, the U.S. helped dissolve the greater social and political move toward democracy which, as Patrick Smith points out, has continued to plague Japan as a country.

The nostalgia that was pervasive in the Tokyo art scene in the 1950s can be seen in relation to the sudden reversal of American policy in Japan at that time. Following WWII and Japan's surrender, two forms of response were evident in the art of the perriod: one revealed a conservative embracing of prewar values; the other was characterized by an enthusiasm over Japan's new beginning. This new beginning was embraced by those in opposition to the imperialistic values that were perceived as having led Japan into the war and to their subsequent defeat. Embracing a more traditional view of art and its forms, the contemporary art scene in Tokyo—which functioned as the country's art capital, with its academies and juried exhibitions—sought to resuscitate prewar values by supporting artists whose work followed traditional art forms and practices, such as calligraphy, Ukiyo-e style woodcut printing, Japanese pottery, and Noh drama. In contrast, the artists of the performance group Gutai looked more toward the West, particularly abstract expressionism, in an attempt to articulate contemporary experience and the desire for a different future. Frustrated with the Tokyo art scene and its piety toward traditions they felt were bankrupt in light of the atrocities of the war, Gutai sought to disrupt this conservatism, to haunt its sentimentality, to jolt its amnesia, through its spectacular performances.

Based in Osaka, the Gutai group formed roughly in 1955 around Yoshihara Jiro, an established painter and leader in the Japanese pre-war avantgarde. Yoshihara's teachings—touched by both the newly discovered abstract expressionism that had been imported through various art journals and international exhibitions, and the frustrated growth of a purely Japanese modernism—deeply influenced the course of Gutai (Munroe 1994).

The word "gutai" literally means "concreteness." This concreteness emphasizes materiality and reveals Gutai's interest in physicality. In their performances (every Gutai work, whether painting or sculpture, was based upon a performative moment) one senses a desperate move toward the world, toward its very fabrication, and, further, toward reestablishing a more direct and tangible tie to art making and its very objectness. Gutai aims to overcome the distance between the artist's touch and the final product. This distance becomes agitated in Gutai work, a move that can also be seen as a political articulation: in stepping across the line Gutai struggles to redefine the limits of individual experience within the social arena.

The work of Gutai can be seen as an aestheticization of physical aggression. Works of art are produced by forcing one's body into contact with a material object or set of objects. This can be seen in the work of Murakami Saburo. For his performance *Many Screens of Paper* (1956), he suspended a series of frames stretched with paper in a row. The artist than ran through the sheets of paper, forcing his arms and legs against the surfaces, thrusting outward against the material. What were left were the marks of this action: a series of ruptured surfaces, broken paintings, "body action-drawings" (Pollock raised to the power of 10). Another Gutai work produced by Shiraga Kazuo, *Challenging Mud* (1955), was a performance piece in which the artist struggled in a circle of mud. Lying in the center of this mass of earth, the artist wrestled against the material, caught in the thickness of the mud, moving against its density. What remained was his expended energy as revealed in the pockets and impressions left in the mud's surface—indexes of struggle.

Others in the group worked with painting and sculpture, using the force of gesture to create abstract fields of intensity and motion. The body was a site of "objectness," as in Tanaka Atsuko's *Electric Dress* (1956), which turned her body into a living sculpture of colored lightbulbs.

What distinguishes Gutai is the simplicity and elegance with which the physicality of materials and actions meet. This collision of body and material—the body of the artist against the body of the world—opens onto a sensuality, a "poetic politics." The effect is a radical lyricism: indexes of struggle and hope, a kind of writing that itself aims to establish a different order. The destruction of objects is necessary for developing this order; in the rending of artistic experience, the acquiescence from which lived experience unfolds is shaken. Because of this, a respect remains for the material object, for its tangibility holds the very promise of change. Collapsing the distance between subject and object sets in motion this potential: in the space where the hand penetrates the object, or pierces paper, or where the body collapses in mud, a relation is formed which, through its sudden appearance, promises change. This material change is also spiritual, for ultimately Gutai sought to release energies of both body and material, artist and object, through willful destruction.

Gutai's manifesto summarizes this attitude: "Gutai Art does not alter the material. Gutai Art imparts life to the material. Gutai Art does not distort the material. In Gutai Art, the human spirit and the material shake hands with each other, but keep their distance" (in Munroe 1994:84). Gutai becomes a kind of cathartic release, aiming to "heal" the very relationship between the individuial and the world—through a magical alteration of subject and object one does not collapse into indistinction but experiences a heightened awareness as a body, as a consciousness.

Gutai in essence was a resistance to the nostalgic regression to an imagined past, embracing instead the democratic spirit that so many Japanese hoped for. Their work bespeaks a desire for a freedom not yet found, and their attack upon materiality itself can be viewed as an outcry against the very fabric of Japanese society—as if by breaking paper or challenging mud, some other reality would present itself.

Like Gutai, Noise-Music hopes to transcend its very own theatre, to turn the moment of performance into something more. However, Noise-Music does not open up onto a utopian horizon of possibility, an infinity; instead, its magic occurs as an occult practice, a subcultural order with its own cosmology.

Keiji Haino, one of the leading figures in the Japanese Noise scene of the past 20 years, describes his relationship to Noise-Music in terms of shamanism:

> I'm offering up my body as a sacrifice. In terms of the relationship between me and the universe, in order to make myself feel better I have to offer myself up to the universe. I believe in the therapeutic properties of music, this is something I've talked about before—how some music makes you feel good, how it physically relaxes your body. What I can't understand is how the people who make that kind of music believe they can heal people without they themselves experiencing any pain. (in Cummings 1996/97:41).

For Haino, to be involved in Noise, in this implosion of terms, this short-circuiting of the distance between order and chaos, structure and spontaneity, preservation and self-destruction, one must sacrifice coherency, that is, the constitution of one's own body. The limits of one's body, of one's psychological organization, fall apart against the carnivalesque extremity of Noise-Music. One breaches the semiotic gap by falling into the void. Yet the void becomes inhabitable, if only for a moment. In this way, Haino approaches his music as a magic, one that allows access to a certain ontological knowledge. Through this shamanistic relationship of musician and audience, the body and the instrument, music and the universe, one reaches the peripheries of consciousness. This, for Haino, is a process of healing. Yet what exactly is in need of healing? What is this "therapy" Haino speaks of? I would venture to say that the carnivalesque nature of Noise-Music offers a necessary chaos to counteract the rules of social behavior—it stirs the mud of stratified codes. This in itself is a form of healing, contributing to a kind of primal health by extending the boundaries of both order and chaos, of what is given and what is imagined.

Noise-Music is by nature nonacoustical, produced through the very means of electronic amplification. By setting electronic signals in motion Noise-Music closes in on itself, eats itself through an obsession with feedback and electronic circuitry. Constructing homemade devices, breaking fabricated ones, and setting instruments against each other, Noise musicians deconstruct the very tools of modern rock music: guitars become metallic reverberating boards, "feedback bodies"; mixing consoles become instruments; cables and cords become sources of voltage hum, interference devices—broken connections + amplified grounding circuit + severed batteries + turn up the volume + plus = electrical intensity = bloody ears = transfiguration =. This electrical equation continually exceeds itself by adopting chaos as its order; Noise-Music strives to lose control by increasing electronic signals, agitating their output. Just as Gutai sought to disrupt the conventions of art and the relationship of artist to material (the audience to the finished product), Noise-Music breaks open the materiality of music by violating its coherency, its fabrication, the very tools of production. The instrument as an extension of the body pulls one into this very violation—or, as Haino describes, the body suffers in the opening of the void.

In this manipulation of the instrument, the coupling of body and instrument into a cyborg union, Noise-Music opens up musical convention, distorts the norm through an obsessive dedication to the "mal-functional," to that which never stays within the bounds of proper use. Noise-Music extends the avantgarde ethos of chaos as order, interference as composition, performance as ritual, artist as engineer. Here, electronic tinkering uncovers a palette of the unwanted, a by-product of electronics. Noise-Music is a kind of "abject art," exploiting all that messy stuff of mind and body, the sick goop of existence that the pure aesthetics of musical composition must clean up, the resolve must resolve.

Noise musicians in Japan are the product of the postwar fraying. The order has been fraying for some time as Smith argues:

> [T]he society the emperor promised never arrived. Meiji [the period of the Japanese industrial revolution of 1868] freed the Japanese from the feudal castes. They could entertain their individual aspirations. But the modern era did not give them the individual liberty to pursue their aspirations. Meiji turned out to be nothing more than a transition from feudal absolutism to absolutism in nineteenth-century form. Japan remained a communal society—closed instead of open, particular instead of universal, a society of individuals who could cultivate no individual values. The contradiction made Japan what it is today—a place of immense but unrealizable dreams, relentless competition, and near-universal frustration. No matter how contemporary we imagine the Japanese to be, the society promised [by Meiji] is the one they still struggle to attain, whose betrayal they seek to redress. (1997:56)

Like Gutai, Noise-Music extends the democratic impulse by forcing self-expression into the hierarchical strata. The cultural rigidity of Japanese society, from the early feudal period ruled by the strict order of samurai culture (a whole elaborate order determining everything from dress to drinking habits, social status to the etiquette of tea making) into the rapid modernization of late 19th century, a progress fueled by loyalty to the emperor—the "godhead" through whom everyone could become a citizen. This historical extension of power and oppression found its culmination in WWII, whose aftermath saw both a lament and a longing for a new democratic beginning.

The new Japanese subculture follows a lineage of dissatisfaction: (1) demonstrations and outbursts in 1960 over the renewal of the AMPO treaty, the postwar security treaty that bound Japan to American security policies and reinforced American dominance of Japanese politics—demonstrations that were ultimately crushed yet which nonetheless expressed a popular unease; (2) a Japanese feminist movement in the 1970s that aimed to open up opportunities for women and to undermine the official domestication of their individuality (which continues today with the 1999 highly controversial sexual harassment trial of the mayor of Osaka, an unprecedented trial leading to the mayor's resignation); and (3) the end-of-the-centuryy economic collapse of the Asian market and subsequent splintering of Japan Inc.'s corporate ethos. This history of dissatisfaction with the governing system that, in Japan, radically dictates one's sense of self, has left its imprint upon recent generations of Japanese who are exploring their own subcultural alternatives, from Noise-Music to techno-culture, bondage clubs to department store groupies. The expressiveness and radicality of these subcultures mirrors the extreme conditions of the society from which they arise. In listening to Japanese Noise-Music one hears not only the audible signs of a possible future, but also the sounds of an extreme response and reaction. As with all chaos, it can be understood in relation to the order it tries to destroy.

Attali theorizes Noise as a precursor for things to come, a kind of prescient register of a future order. In this way, Noise announces what will come, how the dust will settle and reorganize itself, how music will counter dissonance, how order will adjust to subsume chaos.

Noise-Music occurs as musical tension, an order built out of chaos, a chaos constructed out of a social order. It anticipates its own annihilation by embracing it as a musical gesture, constructing itself out of a movement toward chaos, forcing this antithetical relation of order = chaos = order into a special kind of disruption by closing the binary gap, forcing the terms to short-circuit. Through this it reveals the status quo, teases its features into relief.

Noise by nature is a protest. In its excessiveness it stands against the resolve by which order protects itself; it butts up against the other side of language, an antithetical antagonism agitating the semiotic calm, disrupting the circuitry. Language barely holds "Noise" and "Music" as one—its lexiconic output promises its own annihilation; the sequence of +'s falls off into the void. Like the carnivalesque, it implodes. Yet Noise-Music sustains itself—it functions as "music," becoming a kind of logical order in itself, a genre which can be referred to. Its disturbance exists because in the end it knows that music will catch up, that the magic Haino speaks of will eventually become a parody of shamanism, that the broken instrument will soon be manufactured and marketed. Yet it is this continual tension that makes both the institution of music and the thrust of peripheral experiments that much more provocative, and the contrast between the structures of order and the force of chaos complementary.

Note

1. The legacy of Japanese Noise-Music has had a strong impact on experimental music in the United States, affecting and influencing artists and spawning a North American Noise-Music scene, from Los Angeles and San Francisco to Chicago and New York. Many Japanese Noise artists extensively tour North America, as well as Europe, collaborating with local artists and releasing work on local labels, such as Charnel House in San Francisco, Ground Fault in Los Angeles, and RRRecords in Lowell, MA. In essence, Japanese Noise-Music has pointed the way toward methods of expanding sonic experimentation into a highly charged realm based on volume, electronic manipulation, and low-fi aesthetics. At the same time, the extremity of Noise-Music has also opened up the sonic spectrum in general, widening the margins of music experimentation and influencing contemporary sonic arts.

References

Attali, Jacques

1985 *Noise: The Political Economy of Music.* Minneapolis: University of Minnesota Press.

Cummings, Alan

1996/97 "Interview with Keiji Haino." Ardmore, PA: *Halana Magazine.*

Munroe, Alexandra

1994 "To Challenge the Mid-Summer Sun: The Gutai Group." In *Scream against the Sky: Japanese Art after 1945*, edited by Alexandra Munroe, 83–100. New York: Harry N. Abrams.

Smith, Patrick

1997 *Japan: A Reinterpretation.* New York: Vintage Books

Takemitsu, Toru.

1995 *Confronting Silence: Selected Writings.* Berkeley: Fallen Leaf Press.

Selected Discography:

CCCC

1993 *Live Sounds Dopa: Live in USA.* Yokohama, Japan: Endorphine Factory.

Fushitsusha

1998 *Withdrawe, this sable Disclosure ere devot'd.* Quebec: Victo.

Keiji Haino

1992 *Affection.* Tokyo: PSF.
1994 *Beginning and End, Interwoven.* Aachen, Germany: Streamline.
1995 *The Book of Eternity Set Aflame.* Waltham, MA: Forced Exposure.
1995 *The 21st Century Hardy-Guide-Y Man.* Tokyo: PSF.

Hijokaidan

1982 *Gold Rock: Live 1980–1981.* Tokyo: Alchemy Records.
1991 *Windom.* Tokyo: Alchemy Records.

Merzbow

1991/1996 *Music for Bondage Performance*, vols. 1 & 2. Australia: Extreme Records.
1993 *Deus Irae* (with Null). Tokyo: Nux Organization.
1993 *Sleeper Awakes on the Edge of the Abyss* (with Christoph Heemann). Aachen, Germany: Streamline.
1990 *Antimonument.* Gothenburg, Sweden: Art Directe.
1998 *1930.* New York: Tzadik Records.

Kiyoshi Mizutani

1997 *Waterscape.* Bordeaux Cedex, France: Eros.

MSBR

1997 *Live Electronics at Matsuyama* (with Spastic Colon). Tokyo: Flenix.
1997 *Intensification.* Tokyo: MSBR Records.
1998 *1–12* (with Bastard Noise). Tokyo: MSBR Records.
1998 *Collapseland.* Tokyo: MSBR Records.

KK Null

1993 *Absolute Heaven.* Tokyo: Nux Organization.
1994 *Aurora* (with James Plotkin). Nottingham, England: Sentrax.
1995 *Guitar Organism.* San Francisco/Tokyo. Charnel Music/Nux Organization.
1995 *Ultimate Material II.* Kent, England: Fourth Dimension.
1995 *Ultimate Material III.* Memphis, TN: Manifold Records.
1997 *Extasy of Zero-G Sex.* Tokyo: Nux Organization.

More Facts on the Polywave

G.X. Jupitter-Larsen

I was just about to do my regular weekly show, when the transmitter broke down. It wasn't going to get fixed that night, but I stuck around anyway. I recorded the static that was going over the airwaves during the period that my program would have otherwise been on the air. The next week I played these tapes on my show in their entirety; just in case any one had missed the week before. I got a few calls from listeners, as usual. But this time they all had the same thing to tell me. That the static I recorded at the station was different from the static they had heard coming off their radios at home. So on the following week's program I had the station engineer come on and explain, in detail, how a thing like this happens.

I find there's an absurdity to rot and decay. And to communicate this I've taken a light-hearted, if not flippant, attitude towards everything I do. Entropy is the underlying theme for all of my constructs, be they on radio or not. Most of my live performances at clubs & galleries consist of making a mess. Assisted by my performance troupe The Haters, I've performed by tearing up hundreds of books, by smashing numerous sheets of glass, and by setting large trucks on fire. Once in 1993 I used a giant ion-gun to charge an entire audience to 5000 volts. Audience members chased one another around the club giving each other shocks.

As a result of all this, I've developed something of a reputation for wrecking the venues I perform at. Nevertheless, since 1987 one radio station after another has had me come on to do a live on-air presentation. Most of these on-air performances have consisted of the station broadcasting the sounds of me trashing their studio. This live radio-art can last anywhere from five minutes to four hours; with time set aside for station ID. I've performed mostly on college and community stations like KPFA Berkeley, KZSC Santa Cruz, KFJC Los Altos Hills, and KXLU Los Angeles. Pirate stations in Europe such as Radio Alize Paris, and 104.5 Zurich have also had me do my radiophonic specialty for them. Noise collages made from recordings of such shows were featured in the Festival Internacional De Radio Art on the Radio Nacional De España in 1989, 1990, and 1991.

Entropy was also the underlying theme for the radio plays I've done for the ORF program "Kunstradio." With entropy, the outcome is invariably some variation on the hole. My 1992 radio play *Clici-Clic* was composed solely of amplified hole-punching. A contact-mike was mounted on a hand-held hole-punch and recorded one track at a time on a maximal amount of tracks.

The absurdity I find in rot and decay is the fact that biology is based on unstable molecules transferring energy between stable molecules. This very pro-

Jupitter-Larsen starts off a broadcast with cutting cardboard on FREQUENCE 4, Bordeaux France, Oct. 31, 1991

cess, which gives life, simultaneously takes it away. Because the molecules that are doing the energy transferring are unstable, aging occurs. This is irony. And irony is funny. More recent broadcasts of mine have consisted of slowly pushing live microphones into power grinders. Go ahead, laugh. You know you want to.

Voice Tears

The CD

curated by Allen S. Weiss

Voice Tears
Cover art and design by Toni Dove with JJ Gifford
Mastered by Lou Mallozzi at Experimental Sound Studio, Chicago
Printed and pressed by CBC/IS, Chicago

Thanks to Hibou Blanc

1. *Radio Inferno* (1993)
   Excerpt: Cantos I–VI
   Text and concept: Andreas Ammer
   Music: FM Einheit
   Mix: Thomas Stern
   Executive Producers: Herbert Kapfer, Christoph Lindenmeyer
   Directed by Andreas Ammer and FM Einheit
   Produced by Hörspiel Bayerischer Rundfunk/Hessischer Rundfunk
   With: Blixa Bargeld (Dante's Eyes, Dante's Brain, Dante's Mouth), Phil Minton (Vergil, the Guide), Yvonne Duckswoth (Beatrice and Creatures of Hell), Enzo Minarelli (La Divina Commedia), John Peel (The Radio) and Caspar Brötzmann (The Guitar).

2. *2146 Orte* (1995)
   Excerpt
   Kaye Mortley
   Produced by Hörspiel Abteilung, Hessischer Rundfunk, Frankfurt
   Sound Engineer: Helmuth Schick
   First broadcast: 1 November 1995

2146 Orte is an acoustical piece inspired by Jochen Gerz's *2146 Steine, Mahnmal gegen Rassismus*. In 1990 Gerz undertook a clandestine work of art, an invisible monument against racism. With the help of fine arts students, Gerz began nightly to take up various paving stones in the square in front of Saarbrücken Castle, a former gestapo headquarters, and to inscribe on each stone the name of one Jewish cemetery in Germany, before replacing the inscribed stones, face down.

Kaye Mortley: The audio idea came when I saw the catalog. Esther Shalev-Gerz had wanted to make an "invisible book" to tell the story of this "invisible monument," so she told me. Thus, names and fragments of addresses of 2,146 Jewish cemeteries that had existed in Germany at the start of the Third Reich were printed on tracing paper. One could read—or "guess"—through some 10 pages. A type of visual mixing begging (or so it seemed to me) to be translated into sound.

The text of this Hörspiel—the only text—is an alphabetical listing of cemetery names and addresses.

The list is, of course, a literary form (the Bible, Rabelais, Joyce, etc). But for me (and I have worked on other list-based projects) the real power and fascination of the list lies in the fact that it is a type of narration from which the verb/the action is missing. Thus the nouns/names remain both subject/actor *and* object/acted upon: an ambivalence most appropriate in this particular piece.

The actors of the Hörspiel are the voices of all-Germany.

We divided the country into seven main linguistic regions: Hamburg, Leipzig, Berlin, Frankfurt, Stuttgart, Cologne, Munich. We recorded "found" voices reading random pages of list-text. Every fifth page, one page is read by the same seven voices: the chorus, perhaps, as in a Greek tragedy.

The backdrop is the sound of everyday life—trains, subways, boats, cafés, children—woven into a dense tapestry. Everyday life is banal and noisy. Everyday life is also indifferent: it continues (it has, by definition, to continue) while things are destroyed, after they have disappeared.

The interludes are composed of electronically treated stone sounds recorded in a Jewish cemetery in Paris (but who would know?) and the voices of children in a Jewish kindergarten (and, again, who would ever guess?).

3. *Taking Steps* (1993)
   Kathy Kennedy
   Produced at the Banff Centre for the Arts

4. *A Leap of Faith* (1992)
   Excerpt: Part 6 of *Redefining Democracy in America*
   Written, performed, directed, and produced by Jacki Apple and Keith Antar Mason
   Music composed and performed by Eric Cunningham
   Mix by Jacki Apple and Glenn Nishida
   Recording Engineer: Glenn Nishida
   Recorded at Pacifica Studios, Los Angeles
   Funded by the National Endowment for the Arts

In *A Leap of Faith*, a collaboration between Jacki Apple and Keith Antar Mason, a (white) Euro-American woman and a (black) African American man, born in America in the middle of the 20th century on opposite sides of the dividing line, take us on an imaginary journey through time as they wait for the ghost train in the place where our dreams are born and die. They traverse a landscape that reveals the schisms between official history, memory, and experience, as we simultaneously eavesdrop on their private conversations in post-rebellion L.A.

Apple and Mason built on their middle-class urban upbringings and their commonalities as artists with shared social, political, and cultural concerns to honestly explore the differences in their experiences and perceptions as a result of the chasm created by race. The artists' unrehearsed real-time dialog probes the pain, anger, despair, and hope experienced in the aftermath of violent social upheaval as well as their struggle to find a way to move forward on a shared path into the future. The process of making the piece was a healing ritual.

In addition to being a personal response to the conditions, events, and rhetoric surrounding the L.A. "uprisings" of 1992, this work is also an attempt to place such events in a broader historical context. It was constructed as a multidimensional "map" of America viewed as a conceptual place, a virtual place, and a real place, running on three tracks: sometimes parallel, sometimes merging, sometimes crossing. They are represented by three different sonic environments—music, the train, and the weather—each with their own significations.

A Leap of Faith is the final section of *Redefining Democracy in America*, a six-part series that confronts the deep schisms and contradictions of an America in crisis. It was conceived and produced for radio in 1991/92 by Jacki Apple and commissioned by New American Radio.

5. *Casual Workers, Hallucinations, and Appropriate Ghosts* (1991)
   Toni Dove
   Written with Judy Nylon
   Vocal performance by Judy Nylon
   Sound design with Dana McCurdy at Studio P.A.S.S., Harvestworks, Inc.

   Casual Workers, Hallucinations, and Appropriate Ghosts was first released as a segment of the half-hour radio piece for New American Radio entitled *Mesmer — Secrets of the Human Frame* (1991), subsequently used as the soundtrack for the installation *Casual Workers, Hallucinations, and Appropriate Ghosts* sponsored by Creative Time, Inc., and the 42nd Street Urban Development Corp. as part of the exhibition of the 42nd Street Art Project (1994) at Times Square.

6. *Sex Sound Study #1* (1995)
   John Corbett and Terri Kapsalis
   Engineered by Lou Mallozzi at Experimental Sound Studio, Chicago

7. *Ostentatio Vulnerum: a dead language lesson* (remix: 1996)
   Gregory Whitehead

8. *HeadHole* (1995)
9. *Emile Josome Hodinos* (1992)
   Christof Migone

   HeadHole is a synopsis of a new work, *Hole in the Head* (Ohm editions). An earlier version of *Hole in the Head* was commissioned by New Radio and Performing Arts for the New American Radio series.
   Emile Josome Hodinos is part of the *Transpiring Transistor* series. Other parts were published on the *Radio Rethink* CD (Walter Phillips Gallery, Banff, 1994). The full series will be published as part of the forthcoming *Hole in the Head* CD. Hodinos's writings can be found in *Ecrits Bruts* (PUF, 1979; ed. Michel Thévoz) or in translation in *In the Realms of the Unreal* (Four Walls Eight Windows, 1991; ed. John G.H. Oakes).

10. *descends toujours* (1996)
    Julia Loktev

    descends toujours: I know nothing about him really. I know he came there to find a way of getting lost. Whether he found it, I don't know. He aimed to land in a past life, to find himself as former self to import into now. Counting forward, going backward. He fell. Regressed. Egressed. His body opened up and evacuated the growl. He growled relentlessly for over half an hour, growled like a caveman in hell. Growled like a crazy turbopower throat, his voice never losing its force. Click the double-speed switch on the microcassette recorder that captured it, and the growl becomes a baby crying. A double regression—a backflip on the biological and evolutionary timelines in the same breath. Breathe deep. This was once his voice. Raw unprocessed. Now it is mine to keep. But I don't know him really.

11. *Radio (kKkKkKk) Descartes* (1996)
    Excerpt
    Christian A. Herold
    Frame Sound: Matthew Geraci
    Philosophy Consultant: Michael LeCompte
    Sound Consultant: Anna Dembska
    Voices: Andrea Lumm (Descartes's mother), Norbert Bannholzer, Michael LeCompte, Julia Loktev, Washington Square Park (NYC) flaneurs, live audience glosses
    Mix-ins: Orson Welles; Gregory Whitehead; Arnold Schoenberg; Antonin Artaud; random radio programs
    Title kKkontribution: Kate Tarlow Morgan

    Radio (kKkKkKk) Descartes is a piece for solo performer, tape recorder, and recorded voices. The edited *Voice Tears* recording excerpts a live 16-minute version performed at Tisch School of the Arts/NYU in 1995. Mr. Herold most recently performed the piece in Wales at the 1999 Performance Studies International Conference at a panel he chaired entitled "Uneconomic Performance." The text of the piece is published in *3t Tidsskrift for Scenekunst og Teori*, #6 (1999).

12. *Dizzy, not numb* (1995)
    Excerpt
    Written, directed, and produced by Lou Mallozzi
    Narrators: Katy Roderick, Mark Booth, Paula Froehle, Kevin Henry
    Conversation: Terri Kapsalis, John Corbett, Dawn Mallozzi
    Archaeological improvisation: Shanna Linn
    Telephony: Lillian Lennox, Gregory Whitehead
    Eighty violins: Terri Kapsalis, Dan Scanlan
    Bodies in motion: Goat Island performance group (Karen Christopher, Matthew Goulish, Lin Hixson, Greg McCain, Tim McCain)
    Last words: Meenakshi Dash, Bill Talsma
    Vocalizations: Lou Mallozzi
    Recorded and mixed by Lou Mallozzi at Experimental Sound Studio, Chicago

    Dizzy, not numb is an experimental narrative exploring the corporeal body—in motion, in collision, and at rest, in fact and in fiction—a body translated through a number of written, improvised, and conversational linguistic guises.

13. *Manikay: Transmission through Blood* (1996)
    for didjeridu and four shortwave radios
    Andy Haas
    Produced by Andy Haas and Paul Bento at Toben Project Studio, Brooklyn, NY

14. *Four Minutes Is Forty Years* (1995)
    Harri Huhtamäki and Pekka Lappi
    Script: Michelle Constant, Dirk Hartford, and Jicks Jikazana
    Direction: Harri Huhtamäki
    Sound Engineer: Pekka Lappi
    Narrators: Michelle Constant, Dirk Hartford, and Jicks Jikazana
    Producer: Harri Huhtamäki

    Four Minutes Is Forty Years was produced in the radio documentary course in Johannesburg, SA, May 1995, arranged by the Institute for the Advancement of Journalism, and conducted by Harri Huhtamäki. The course was supported by Friedrich Ebert Stiftung, the South African Broadcasting Corporation, and Radio Atelier of the Finnish Broadcasting Company.

Biographies

Andreas Ammer, born in Munich, Germany, is a freelance author. Since 1993 he has collaborated with FM Einheit and Ulrike Haage. They have won many international prizes for audio art, including the Prix Italia, Prix Futura, Morishige Award (Japan), Gold Medal at the New York Radio Festival (USA), and Hörspielpreis der Kriegsblinden (Germany). *Radio Inferno* was first published in 1993 as a Hörspiel and was issued as a CD (RTD 197.1598.2 42 EGO 203) the same year.

Jacki Apple is a visual, performance, and media artist, audio composer, writer, director, and producer whose interdisciplinary works have been performed, exhibited, and broadcast internationally, featured in festivals, on numerous anthology CDs, and at <www.somewhere.org>. A major retrospective of her audio/radio work 1979–1997 was presented at the international SoundCulture '99 festival in Auckland, NZ. Her CDs include *Thank You for Flying American, ghost.dances, eco-geographies, L.A.Noir,* and *Star Tripping*. She was the producer/host of *Soundings,* a weekly radio show, KPFK-FM, Los Angeles, 1982–1995. She is a core faculty member at Art Center College of Design, Pasadena, CA.

John Corbett is a writer, producer, and musician who lives in Chicago. He has released CDs as an improvisor and audio artist, the most recent of which is *I'm Sick About My Hat* (Atavistic Records), and he curates the Unheard Music Series. Corbett writes for various periodicals including *Down Beat* and the *Chicago Reader*, and he has composed liner notes for more than 100 CDs. His book, *Extended Play* (Duke), was published in 1994. Corbett programs a weekly jazz series and an annual improvised music festival at the Empty Bottle, Chicago. He is Adjunct Assistant Professor at the School of the Art Institute of Chicago.

Toni Dove is an artist who works primarily with electronic media, including virtual reality and interactive video installations that engage viewers in responsive and immersive narrative environments. Her work has been presented in the United States, Europe, and Canada, as well as in print and on radio and television. Her most recent interactvie movie installation, *Artificial Changelings*, uses motion sensing to allow a viewer standing in front of a screen to move a video character's body and generate speech and music. Her current project under development is *Spectropia*, an interactive supernatural thriller.

René Farabet has been producer of the Atelier de Création Radiophonique at France Culture of Radio France since 1969. He has received the Prix Italia (1971, 1998), the Prix Ondas (1982) and the Prix Futura (1985, 1987, 1993) for his radiophonic work. He received his doctorate from the Sorbonne.

Richard Foreman has been a MacArthur Fellow, has received nine OBIE awards and an award for Lifetime Achievement in Theatre from the National Endowment for the Arts. He has written and/or directed and designed over 60 plays around the world for major theatres and festivals and for his own Ontological-Hysteric Theater. Six books of his plays and essays have been published as well as one novel.

Rev. Dwight Frizzell and *Jay Mandeville*, collaborative artists who say they share a third mind, have written numerous articles, essays, interviews, and plays over the past three decades. Their *From Ark to Microchip* radio shows, *Indeterminate Moments with John Cage* and *Contacting the Other: Amazing Psychotropic Tales* are available from LodesTone Media. Their writings on the "Early Radio Big Wigs" were published in *Radiotext(e)* [Semiotext(e), 1993].

Andy Haas is a self-taught didjeridu/saxophone player living and working in New York City. He has performed in Canada, Europe, Japan, the United Kingdom, and the United States. His CD of didjeridu duets entitled *Arnhem Land* is available on the Japanese Avant label, and he is currently working on a new recording for the Tzadik label to be released in 2001. "Alef-Beit," for didjeridu and recitation of the Hebrew alphabet, is available on the Knitting Factory compilation CD *A Guide for the Perplexed*. He has performed live with radio and television transmissions for some 20 years.

Christian A. Herold is a writer, performer, and adjunct professor living in Greenwich Village. His verse drama *Multiple Play* was produced at the 13th Street Repertory Theatre; his play *Antilogic* awaits a producer. He has performed other of his works in Washington, D.C., Boston, New York, and Vermont. He is a Ph.D. candidate in Performance Studies at Tisch School of the Arts/NYU. His master's thesis studied screaming and his dissertation is about talking-to-oneself in Beckett, Cage, Ashley, and Lucier. He's on the editorial board of *Women & Performance* and teaches in NYU's Drama Department.

Mary Louise Hill holds a Ph.D. in Performance Studies from Tisch School of the Arts/NYU, and "Developing a Blind Understanding" was part of her dissertation for that degree. Other material from that work has been published in *Women & Performance* or presented at conferences in both the United States and Turkey. Currently a lecturer in American Culture and Literature at Baskent University in Ankara, Turkey, she continues her research in sound, technology, and gender, while pursuing new research in colonialism and the history of Cyprus. Her current work-in-progress is a novel, entitled *Past Remedies.*

Harri Huhtamäki, born in 1949, is a Master of Art and theatre director. He is the author of numerous radio documentaries, features, and radio plays as well as *The Five Ways of the Radio* (Like-Publishing House, Helsinki, 1993). He has won the Berlin Prix Futura first prize for *Cockroach* (1989), the Pohjanmaa National Critics' Award for *Rahkonen* (1990), and the Nordic Media Prize for *Requiem for Nordic Trees* (1993). Huhtamäki has conducted radio workshops in Australia, China, Fiji, Hungary, the United States, South Africa, and Brazil, and works at present as head of Radio Atelier of the Finnish Broadcasting Company.

G.X. Jupitter-Larsen is a performance artist and noisician with over 225 performances, both on stage and radio, as well as over 40 record and CD releases to his credit. Best known for trashing clubs and recording what are considered the noisiest releases ever, Jupitter-Larsen still continues to come up with new and different ways to make a mess: "Happy fun time for the entropically literate." Recent shows have consisted of performers using power grinders to wear away at the car tires that each had over his or her shoulders. Records and CDs by Jupitter-Larsen consist of noise collages made mostly from the sounds of things falling apart.

Douglas Kahn, Associate Professor of Media Arts at the University of Technology, Sydney, has written *Noise, Water, Meat: A History of Sound in the Arts*

(MIT Press, 1999) and coedited *Wireless Imagination: Sound, Radio and the Avant-garde* (MIT Press, 1992/2001). With Daniel Tiffany and Karen Pinkus, he edits the *Auditory Cultures* book series from MIT Press, and is an international editor for *Leonardo Music Journal*. In collaboration with Frances Dyson he works under the name Liminal Product on media production and on writing and speaking on contemporary arts, music, media, culture, and politics.

Terri Kapsalis is a writer, performer, and improvising violinist based in Chicago. Her writings have appeared in such venues as *Lusitania, New Formations, Public, The Baffler*, and the *Chicago Reader*, and she is the author of *Public Privates: Performing Gynecology from Both Ends of the Speculum* (Duke University Press, 1997). Kapsalis is a founding member of Theater Oobleck and can be heard on a number of CDs, including Tony Conrad's *Slapping Pythagoras*, Gastr Del Sol's *Harp Factory on Lake Street*, and with John Corbett and Hal Rammel on *Van's Peppy Syncopators*. She teaches at the School of the Art Institute of Chicago.

Alexandra L.M. Keller, novelist and cultural critic, is a writer and film scholar; she received her Ph.D. in Cinema Studies from Tisch School of the Arts/NYU.

Kathy Kennedy is a sound artist with a background in classical singing. Her practice generally involves the voice and issues of the interface of technology, often using telephony or radio. She is also involved in community art and is a founder of the Digital Media Resource Center for women in Canada, Studio XX, as well as choral groups Choeur Maha in Montreal and ESTHER in San Francisco. Her large-scale sound installations, called sonic choreographies, have been performed at Lincoln Center's Out of Doors Series, at the inauguration of the Vancouver New Public Library, and at Place des Arts in Montreal.

Brandon LaBelle is an artist and writer from Los Angeles. Through installation and performance his work draws attention to the phenomenal dynamics of found sound. He was recently featured in the *Sound As Media* exhibition at ICC, Tokyo. He is a writer of essays and creative fiction, addressing issues pertaining to sound-art, architecture, and the poetics of experience. He is the co-editor of *Site of Sound: Of Architecture and the Ear*, published by Errant Bodies Press.

Julia Loktev's work started with live radio, which led to produced audio pieces, which began to incorporate visual spaces, which led to a desire to populate those spaces with characters, which led to installation, to video, and to film. Her first feature film, *Moment of Impact* won a Directing Award at the Sundance Film Festival and has been shown in over three dozen international film festivals including MOMA's New Directors/New Films and the San Francisco International Film Festival. She was born in St. Petersburg, Russia, and lives in New York.

Lou Mallozzi is an audio artist from Chicago who dismembers and reconstitutes language, sound, and gesture on radio, recordings, stages, and sites. His radio works have been broadcast in North America, Europe, and Australia. He has presented live sound performances and improvised music in individual and collaborative projects, and has presented sound installations at several galleries. He is the Director of Experimental Sound Studio in Chicago and teaches at the School of the Art Institute of Chicago.

Keith Antar Mason is a poet, playwright, performance artist, and director and founder of Hittite Productions, a black male performance/theatre collective whose outspoken, often confrontational works have been presented at major venues throughout the United States. His radio piece *Frenzy in the Night*, produced by Jacki Apple, was commissioned by New American Radio.

Christof Migone is an audio and performance artist whose work has been broadcast and exhibited internationally. He holds an M.F.A. from the Nova Scotia College of Art & Design and is currently completing a Ph.D. at NYU's Tisch School of the Arts/NYU's Department of Performance Studies. His writings have appeared in *Musicworks*, *Semiotext(e)* (1993), *Radio Rethink* (1994), *Cahiers Folie Culture*, *TDR*, *Angelaki*, and *XCP: Cross-Cultural Poetics*. He is the coeditor of *Writing Aloud: The Sonics of Language* (Errant Bodies Press, 2001).

Joe Milutis is a writer and media artist whose work on electronic art has appeared in *Afterimage*, *Artbyte*, *Soundsite*, and *Wide Angle*. He is completing his dissertation "Administering the Ether, and the Aesthetic of the Absolute" for the Modern Studies Program in the Department of English at the University of Wisconsin-Milwaukee.

Kaye Mortley has been an independent radio producer since 1981. Based in Paris, she often works for the Atelier de la Création Radiophonique at France Culture, as well as for other European broadcasting organizations, and for the A.B.C. in Australia. She has been awarded the Prix Futura (1981, 1985, 1987) and the Prix Europa (1998).

Fred Moten was born and raised in Las Vegas, Nev., and now lives in New York City where he is Assistant Professor of Performance Studies in the Tisch School of the Arts/NYU. Moten has published numerous scholarly articles and is currently at work on a project called "Animaterial (Some Black Performances)" which focuses on the politics of sound in black performance. He has also published poetry in *Grand Street* and *Lift*, and has poems forthcoming in *Callaloo* and *Five Fingers Review*. His first chapbook, *Arkansas* (2000), was published by Pressed Wafer Press.

Mark S. Roberts teaches philosophy at SUNY/Stony Brook. He has published numerous articles in the fields of continental philosophy, aesthetics, psychoanalysis, and media theory, and has edited seven books in philosophy, psychoanalytic theory, and cultural studies. His most recent works are *Disordered Mother or Disordered Diagnosis? Munchausen by Proxy Syndrome* (with David B. Allison, The Analytic Press, 1998), and *High Culture: Reflections on Addiction and Modernity* (with Anna Alexander, The Analytic Press, forthcoming).

Susan Stone is a producer, writer, and actor, and Director of Pacifica Radio/KPFA-FM's Drama and Literature Department (Berkeley, Cal.), where she is also executive producer of the weekday documentary series *Audio Evidence* and weeknight world arts series *The Eleventh Hour*. Her independent productions of original mixed-media texts and sound design for theatre, film, and television have received awards from the National Federation of Community Broadcasters, the San Francisco Film Festival, and American Women in Radio and Television.

Allen S. Weiss has written and edited over 25 books, including *The Aesthetics of Excess* (SUNY Press, 1989); *Perverse Desire and the Ambiguous Icon* (SUNY Press, 1994); *Phantasmic Radio* (Duke University Press, 1995); *Sade and the Narrative of Transgression* (Cambridge University Press, 1995); *Taste, Nostalgia* (Lusitania, 1997); *Mirrors of Infinity* (Princeton Architectural Press, 1995); and

Unnatural Horizons (Princeton Architectural Press, 1998). He directed *Theater of the Ears,* a play for electronic marionette and taped voice based on the writings of Valère Novarina. He teaches in the Departments of Performance Studies and Cinema Studies at Tisch School of the Arts/NYU.

Gregory Whitehead is a playwright, voice performer, and international radio artist who has produced over 100 radio plays, voiceworks and documentary features, including *Shake, Rattle and Roll* (Prix Futura), *Pressures of the Unspeakable* (Prix Italia), *L'Indomptable* (with Allen S. Weiss for France Culture), and his most recent play, *The Marilyn Room* (BBC Radio 4). He is also the coeditor of *Wireless Imagination: Sound, Radio and the Avant-Garde* (MIT Press, 1992/2001). "Radio Play Is No Place" was originally published in the Grenoble-based magazine *Revue & Corrigée.*

Ellen Zweig works with text, audio, video, performance, and installation. She uses optics to create camera obscuras and miniature projected illusions. She has presented work in Europe, Australia, and the United States. As an artist-in-residence in the Interactive Telecommunications Program at Tisch School of the Arts/NYU, she created a collaborative performance over Internet2 with MIT. Now in-residence at MIT, she is planning a serial performative narrative on the internet. Among other projects is the recently completed novel, *Mendicant Erotics*, which began as the radio-play in this volume.

Index